The Institute of Biology's
Studies in Biology no. 61

The Biology of Eucalypts

Lindsay D. Pryor
D.Sc.
Professor of Botany,
Australian National University, Canberra

Edward Arnold

First published 1976
by Edward Arnold (Publishers) Limited
25 Hill Street, London W1X 8LL

Board edition ISBN: 0 7131 2542 x
Paper edition ISBN: 0 7131 2543 8

Printed in Great Britain by
The Camelot Press Ltd, Southampton

General Preface to the Series

It is no longer possible for one textbook to cover the whole field of Biology and to remain sufficiently up to date. At the same time teachers and students at school, college or university need to keep abreast or recent trends and know where the significant developments are taking place.

To meet the need for this progressive approach the Institute of Biology has for some years sponsored this series of booklets dealing with subjects specially selected by a panel of editors. The enthusiastic acceptance of the series by teachers and students at school, college and university shows the usefulness of the books in providing a clear and up-to-date coverage of topics, particularly in areas of research and changing views.

Among features of the series are the attention given to methods, the inclusion of a selected list of books for further reading and, wherever possible, suggestions for practical work.

Readers' comments will be welcomed by the author or the Education Office of the Institute.

1975

The Institute of Biology,
41 Queens Gate,
London, SW7 5HU

Preface

This booklet is intended to provide an introduction to the biology of the eucalypts. These broad-leaved evergreen plants are like other woody dicotyledons in some general biological features, but they differ in various ways by specialization in morphology, physiology, ecology and genetic aspects. They are confined in natural occurrence entirely to the Australasian region.

The emphasis recently on the widespread need for increased wood production, especially in warm temperate and tropical areas in developing countries, has made the genus *Eucalyptus* recently one of the most widely planted silvicultural crops. It is used also on a large scale to provide a short rotation crop yielding wood for industrial use. Perhaps this will increase with growing energy needs.

Thus the biology of the group will interest many people because of its importance for plantation silviculture and amenity planting in many places throughout the world. Some of the special biological aspects of the genus are a rewarding study simply because they are unique.

Canberra, 1975

L. D. P.

Contents

1 Geographic Distribution of Eucalyptus

1.1 Distribution beyond Australia

Eucalyptus is very much an Australian specialty, for almost all of the 600 or so species and varieties are endemic to the country and a small part of Papua, in Daru and near Port Moresby, where there are about five or six species which are also shared with Australia. These species are common in Australia's Cape York Peninsula and other parts of northern Australia.

Two species occur also in Timor and some adjoining islands of the Lesser Sunda group Flores, Solor, Alor and Wetar. One of these two species is in the group found in Papua. The other is different, since it is not known to occur naturally in Australia nor has it been found so far in New Guinea or West Irian. It is therefore a species endemic apparently to the Timor area. There is as well a species, *Eucalyptus deglupta*, growing

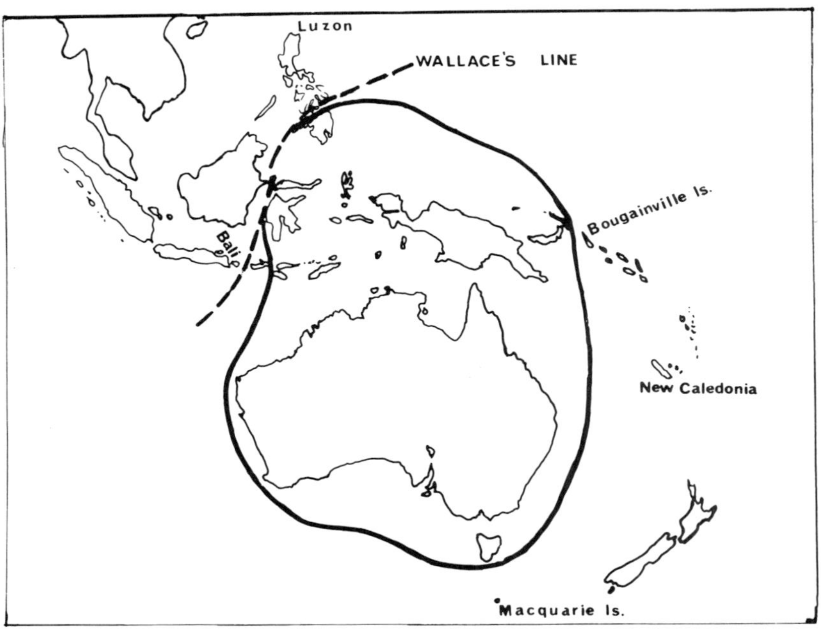

Fig. 1.1 The limits of natural distribution of *Eucalyptus* in the Australasian region (solid line) showing its close conformity on the north-west boundary with Wallace's line which it does not cross to the Asian side.

naturally in New Britain, northern New Guinea, Sulawesi and Mindanao which is likewise endemic to these islands and does not occur in Australia (Fig. 1–1). *E. deglupta* is quite unlike any other Australian eucalypt, while the species from Timor is evidently related to one of eastern Cape York Peninsula although it is certainly distinct from it.

Because of the great diversity of species in Australia it is assumed that the genus has had this area as its centre of development and that in the evolutionary past the progenitors of *Eucalyptus* were present here and then gave rise in time to the great array of species now existing.

It is assumed also that species (excluding *E. deglupta*) in Papua New Guinea and in the Timor area migrated there at some time from the Australian centre, but the route to Timor is obscure, unless it were through West Irian, where so far no trace of it or its relatives has been reported.

It is interesting also to note that *Eucalyptus* does not occur naturally either in New Caledonia or New Zealand, in spite of the fact that there are, on the one hand, a number of similarities between New Caledonian plants and Australian groups, for example, in *Melaleuca* and *Boronia* and, on the other, between New Zealand and Australia, for example in *Leptospermum*. The genera *Melaleuca* and *Leptospermum* are not only in the same botanical family as *Eucalyptus*, that is the *Myrtaceae*, but they are also in the same subdivision of that family, the sub-family *Leptospermoideae* which has capsular fruits.

1.2 Distribution within Australia

The most southerly occurrence of eucalypts is in southern Tasmania in about latitude 43° S. Beyond this there is little land before Antarctica and the speck that is Macquarie Island in latitude 55° S., which is politically part of Tasmania and therefore in that sense Australia's southern-most part, is without woody plants and certainly has no eucalypts. The species in Tasmania are for the most part also on the Australian mainland. This gives no cause for biogeographical surprise, since Bass Strait is shallow enough to allow a land connection between Tasmania and Victoria during periods of sea-level fall which without doubt occurred during the last and probably earlier glacial periods.

There are, however, several species endemic in Tasmania which do not appear at all on the mainland. This occurrence of endemics of limited geographic distribution seems to be independent of the present-day Bass Strait and there is no need to involve it as an isolating factor necessary to lead to such speciation but rather to attribute the endemism to the environmental features of Tasmania which, because of its more southerly latitudes, is unique in the Australian region.

On the mainland eucalypts are widespread but they are not common in the desert zone (Fig. 1–2), there being confined mostly to special sites such

Fig. 1.2 The areas (stippled) in Australia in which *Eucalyptus* is prominent in the vegetation. It is not entirely absent from the unshaded areas, especially in the arid central and central western regions, but when it does occur it is confined to limited sites such as seasonal water courses or rock clefts. These occurrences cannot be represented at the scale of the diagram. (Based on field map by J. A. Carnahan)

as water-courses, rock clefts or particular substrates—ecological islands as it were—often well separated from each other with the intervening areas occupied by species of other kinds of plants including especially *Acacia*. But the better watered parts of the country relative to the central and north-west arid zone are the home of the eucalypt, and it is in this zone also that most Australian people live. The exceptions to the otherwise ubiquitous eucalypt in these more populated parts are two. Firstly, there are disjunct patches of Indo-Malayan type rain-forest from Cape York down through eastern Queensland and New South Wales to the Victorian border. This vegetation type is closed to *Eucalyptus* and providing it is not changed in some way as by firing, the eucalypt is excluded from it completely. Secondly, in western Tasmania and in a few southern Victorian localities, temperate rain-forest reminiscent of those in New Zealand and Chile and consisting of *Nothofagus* holds sway. This also cannot be penetrated by eucalypts unless there is some damaging or similarly persisting deleterious factor.

The Indo-Malayan rain forest of northern Queensland and the east

Australian coast is the vegetation type which luxuriates in the east–west oriented mountains of Papua New Guinea and West Irian and occurs there for the most part even down to sea-level. This belt seems to have formed an uncrossable barrier for almost all eucalypts. Torres Strait, between Cape York and Papua is quite shallow and is not to be thought of as a barrier to plant migration, being both in distance and depth even less significant than Bass Strait in the south.

Within Australia, the eucalypts are distributed largely according to climatic environment with various qualifying additional factors, especially soil type. Perhaps the major difference which is largely independent of climate is that between south-western Western Australia and south-eastern Australia. In both areas the rainfall is predominantly in winter and the temperature regimes are comparable. Also the main soil types are similar. But there are no species shared between the two regions, in spite of the fact that evident though relatively remote affinities can be seen in various species pairs. There is nowadays a climatic and soil zone—the Nullarbor Plain—which completely separates the two and the few species which straddle the Nullarbor are limited to the more arid fringes or special coastal sites and do not extend to the general regions referred to.

Between the tropical north and the temperate south, while there is not much sharing of species, on the east coast the eucalypt zone is continuous with species occupying their characteristic geographic range, each at their southern or northern limits to be replaced by another. However, the affinity of some species in the extreme north with those found well south is in several cases much closer than between those of the south-west—Western Australia—and south-eastern zones in spite of the fact that there are gross climatic differences between the tropical north with its summer rainfall and the temperate south with its largely winter rainfall and much colder winters. Indeed, one species, *E. tereticornis* occurs along the whole of the east coast of Australia as well as in Papua and has its southern limit in south-eastern Victoria on Bass Strait. *E. camaldulensis* has almost the same latitudinal spread but it does not reach Papua. Thus it seems that the south-west of Western Australia and the south-east of Australia in N.S.W., Victoria and Tasmania have been long isolated botanically although the plants of both areas must have had a common, though ancient, origin.

2 Morphology

2.1 General

The eucalypts are evergreen woody plants which are mostly trees and many are very large. The Mountain Ash, *E. regnans* of Victoria and Tasmania (Fig. 2–1) reaches about 100 m in height and therefore (for what this comment is worth) is the tallest species of broad-leaved tree in existence. Several species reach more than 70 m, as Karri, *E. diversicolor* in Western Australia, Flooded Gum, *E. grandis* in New South Wales, and Kamarere, *E. deglupta* in New Britain. Some are merely shrubs, such as *E. moorei* of the Blue Mountains of New South Wales, or *E. vernicosa* of high altitude sites in Tasmania, or *E. tetraptera* of the coastal sand plain areas of Western Australia.

However, most species are forest trees of 30–50 m height or woodland trees of 10–25 m in height, while thirty to forty species have a distinctive form called Mallee which is characterized by having several stems from a common underground woody stock (Fig. 2–2).

2.2 Adult leaves

A most distinctive vegetative feature is the adult foliage. The general type which is found in the majority of species is a petiolate leaf, falcate-lanceolate in shape and much the same in appearance on both surfaces, that is to say, it is isobilateral. Coupled with this the leaves hang on the branches in a more or less vertical alignment so that looking upwards the crown of most eucalypts appears thin and casts only light shade at midday, whereas side shade in morning or afternoon may be relatively heavier. Species which differ from this condition are in a minority and often attract comment. In some species the leaves are dorsiventral and evidently bicolorous and as well are hypostomatous, with stomata only on the lower, abaxial surface as is general in most broad-leaved plants. But whatever variation occurs in shape and form all are leathery and tough textured which earns them the name Sclerophyll which is widely used in describing in a general way the Australian eucalypt vegetation (PENFOLD and WILLIS, 1961). Sclerophyll vegetation (on a world basis) is composed of many genera and species as well as *Eucalyptus*, but most of the Australian Sclerophyll vegetation is characterized by having as its tree component various species of *Eucalyptus* with *Acacia* and a few other genera completing the array.

Fig. 2.1 *Eucalyptus regnans*, Mountain Ash, forming wet sclerophyll forest in Tasmania. It is the tallest eucalypt and the tallest broadleaved plant.

Fig. 2.2 Mallee in western New South Wales. The species are *E. oleosa* and *E. dumosa*.

2.3 Juvenile leaves

While there are differences in adult leaf morphology as described above, the pervading overall sameness of the mature foliage is the impressive feature of the genus rather than the departures from this standard. For the juvenile foliage it is different. Many eucalypts are markedly heteroblastic—that is to say they have leafy shoots of more than one type which differ markedly from each other (Fig. 2–3). These are usually referred to as juvenile and adult phases and the leaves they bear as juvenile or adult leaves. The juvenile type leaf may be radically different from the adult. One of the best known is *E. globulus* in which the juvenile

Fig. 2.3 A leafy shoot of *Eucalyptus* showing a rapid transition from A. juvenile foliage; B. Adult foliage.

leaves are opposite, sessile, highly glaucous, oblong-acuminate in shape and dorsiventral while the adult leaves are alternate, petiolate, non-glaucous, falcate-lanceolate and isobilateral.

The difference is so striking that when a single tree is carrying both juvenile and adult leaves, as all do at some phase in their development from seedling to adult, and the contrast is at a maximum, the question is often put as to what grafting programme has been involved since the two types of leaf are so different.

In descriptions and general use, the term juvenile leaf refers to those which occur on seedlings 2–3 feet tall. The first few pairs of leaves in a young seedling where the cotyledons are still present may be called seedling leaves and may be rather different again from either juvenile or adult leaves but are infrequently included in diagnostic descriptions. Those which occur in the phase of growth where successive leaves show a morphological transition from juvenile to adult may be referred to as intermediate leaves.

The juvenile leaves are of interest from several viewpoints. They are particularly useful in precise species diagnosis. There are numerous pairs of species which are closely similar in adult leaf and floral characters but which differ extremely in juvenile leaf morphology as for example in *E. viminalis* and *E. dalrympleana*. Thus if juvenile foliage can be seen the discrimination between specimens of these two species is precise and simple.

Juvenile leaf characters are also known to be strongly inherited and since they can be seen in plants a few months after seed sowing, access is gained to features which are powerful markers in genetic analyses. The physiological characteristics and adaptive value of juvenile leaves are also related to various aspects of evolution and speciation in the genus.

Not all species are markedly heteroblastic but most are in some degree characterized in this way and the feature is always included in species descriptions. Juvenile leaves are seen first in young plants and are usually fully expressed in the morphology of the leaves developed at the 10th node on the stem beyond the cotyledons. Leaves produced at subsequent nodes beyond the 10th may continue to display juvenile morphology and in some species this may persist for very many nodes. A plant may be three or four years old and perhaps 7 m tall but still with juvenile foliage. With most, however, after about the 20th or 30th node and 2 m of height, leaves formed at successive nodes go through a series of shape differences occupying (in sequence) a number of nodal positions until finally, fully adult leaves are produced. The number of nodes in which the transition occurs varies with species. Also the absolute node number at which the fully adult leaf first appears differs with different species as well as to a lesser extent in different individuals of the same species. In a few, under natural conditions at least, the adult type leaves never develop and so the tree though fully mature in an ontogenetical sense still has juvenile type leaves. This is seen for example, in *E. cinerea*, *E. melanophloia*, *E. setosa* and *E. risdonii*—species in widely separated systematic groups through the genus. The places in which these species grow and their systematic grouping does not allow any easy deduction as to the reason why some species though successful, differ in this morphological character from the majority. It is a feature, of course, which is often valued horticulturally and such species are often planted for their ornamental value.

2.4 Lignotubers

An unusual feature seen in *Eucalyptus* is the presence in most species of lignotubers (Fig. 2–4). These organs though especially characteristic of

Fig. 2.4 Lignotubers: A. on a seedling of *E. bloxsomei* and B. *E. grandis*, Flooded Gum without tubers.

the genus are not confined to *Eucalyptus* but do occur in some other *Myrtaceae* and perhaps in still other groups. They are first seen in most eucalypt species during the first year of life of the plant as two globular swellings in the axils of the cotyledons. Additional pairs of tubers may occur in the axils of the first and second and even the third pairs of seedling leaves. With time, they grow and if there are more than the cotyledonary pair they coalesce to form a quite large woody tuber up to several centimetres in diameter. If the plant survives and continues to grow, differential stem increment finally overtakes the tuber and it is no longer to be detected in most species by the time the stem is 20 centimetres in diameter. The lignotuber has been shown to consist of a mass of vegetative buds and associated vascular tissue and to contain substantial food reserves. If the top of a seedling which has developed a lignotuber is for some reason destroyed as by fire, drought or grazing, growth is vigorously renewed by the development of new shoots from the lignotuber. It is evident that this is an organ of very considerable survival value in an environment where fire and drought are frequent as well as grazing. It is therefore of interest to know that there are several species, perhaps thirty to forty, which have not been seen to develop lignotubers. Species which develop lignotubers usually do so in almost all individuals. The absence of a lignotuber is associated with particular behaviour. Such species are generally prolific seed producers and if damaged by some agent, such as fire, to which these species are often fairly susceptible, the regeneration which follows is extremely heavy. Thus it is that often in species where the lignotuber does not develop, an alternative survival mechanism has emerged which confers in an alternative way an advantage to the species in survival capacity.

2.5 Eucalyptus Buds

The bud system of the leafy shoot of *Eucalyptus* has some unusual features (JACOBS, 1955). In the axil of each leaf there are always originally two buds, one is called a naked bud in that it has no bud scales and is evident always in the axil leaf. It is the source during an ordinary growing season of the lateral branches which go to make up the divaricate leafy shoot. At the same time there is a second bud which is covered by petiolar and axillary tissue and is dormant. This is not evident unless exposed by a longitudinal section. In the first growing season it does not develop ordinarily to form a leafy shoot but in many natural eucalypt environments, the stresses of the dry season, or in the southern parts of Australia, of winter, lead to a cessation of growth and frequently to aborting and death of the naked buds. Thus when growth resumes in the following season, it is usual for the new shoots to be produced from concealed buds. During the ontogeny of any individual not all concealed buds are stimulated to growth after a season of stress and some of them remain dormant indefinitely so far as producing leaves is concerned, slowly extending out just under the bark surface as the stem or trunk increases in thickness. Their presence can be detected by a depression in the stem but their existence is appreciated most readily when an individual tree is subjected to defoliation either by fire, drought, leaf-eating insects or mechanical damage. Then the dormant buds are

Fig. 2.5 Wet sclerophyll forest of *E. fastigata* and *E. viminalis* about 10 months after a heavy fire.

released from their resting state and they sprout to form leafy shoots, called epicormic shoots, all over the trunk and main branches. In this event the morphology of the leaves produced from dormant buds is of the juvenile type and a very common after-fire feature in eucalypts is a period in the regrowth year when the leafy crown is made up of shoots derived from this source (Figs. 2–5, 2–6).

Fig. 2.6 Epicormic shoots on the main trunk of *E. fastigata* after fire.

2.6 Phyllotaxis

Leaf arrangement in *Eucalyptus* follows an interesting pattern. In the adult stage *Eucalyptus* leaves are alternate and might be thought to have one of the spiral phyllotactic arrangements, common in many broad-leaved plants. However, the alternate condition in the genus has developed as a modification of what is the basic phyllotactic arrangement which is taken to be that displayed by the juvenile leaves in which the placement is opposite and decussate (JACOBS, 1955). This means that succeeding pairs are oriented at right angles to the foregoing leaf pair. In the juvenile state the leaf pairs are separated by a length of shoot which is simply termed the internode but when the adult type leaf develops there is in most cases a separation of the two members of each leaf pair by a length of stem which has been designated the intranode. With a little care one can deduce which individual leaves should be paired and, by untwisting any rotations which develop in the stem (which is usual), it will be seen that the leaf pairs are now no longer opposite but alternate, because the points of petiolar insertion are separated by the intranodal stem tissue. However, they are in fact still decussately arranged. Thus *Eucalyptus* has achieved the distribution of its foliage by a modification of the decussate pattern (BROOKER, 1968). This is in contrast with the more common spiral type of phyllotaxis which is found in many broad-leaved flowering plants.

2.7 Stem features

Eucalyptus is notable for the high proportion of species which have decorticating bark which results in smooth stems often of white or light coloured appearance (CHATTAWAY, 1953). A similar habit is found in unrelated species of broad-leaved plants such as *Platanus* (Sycamore) or *Terminalia arjun* (Arjuna) and even in pines such as *Pinus bungeana* but seldom are so many species in a single genus found with this feature. Most tree genera have persistent dead outer bark like the remaining eucalypts. Bark-shed in decorticating species is seasonal and often associated with a colour change—*E. rubida* for example is named because of the rosy colour which the trunk assumes a little before the dead outer bark is shed in midsummer. The bark may shed in a patchy way to give a mottled pattern as in *E. maculata*, Spotted Gum, or the beautiful pastel shades of Kamarere, *E. deglupta*.

2.8 Floral morphology

The traditional way of classifying flowering plants is to give emphasis to the structure of the flowers and to assess affinity largely by floral similarities. This approach, of course, has not been applied exclusively and in practical terms, both vegetative and floral features are taken into consideration in making classifications. However, in the broader aspects, more reliance has been placed on floral morphology and this is still used in such assessments. The most distinctive feature of the *Eucalyptus* flower is the presence of an operculum (PRYOR and KNOX, 1971). This, before anthesis, is a cap covering the reproductive organs. The cap is shed at anthesis and the vital reproductive processes subsequently are set in motion. The operculum in most eucalypt species is actually a combination of two such caps, outer and inner. These are interpreted as being calyx and corolla respectively of a tetramerous flower in which the four separate sepals are united into a single cap and the separate petals into the second inner cap. Only the tips of the sepals and petals at the most, remain discrete leaving a tiny canal between them. Robert Brown who travelled with Matthew Flinders in exploration in Australia in 1805, interpreted these structures in this way and there has been no need, in most species, to regard them differently since his time. But not all species have the double operculum, the main differences being in a relatively few in which the calyx remains composed of separate sepals which appear as teeth at the top of the receptacle (hypanthium) within which whorl there is an operculum composed evidently of four fused petals, as for example in *E. tetragona*. Robert Brown was first also to offer this interpretation of this group of species. In still another group there are no sepals to be seen as separate teeth and there is only one operculum evidently visible (CARR and CARR, 1959b). This cannot be interpreted readily as either calyx or corolla nor even as a fusion of both and differing views are held about the nature of the structure. The *Eucalyptus* flower has some of the general Myrtaceous

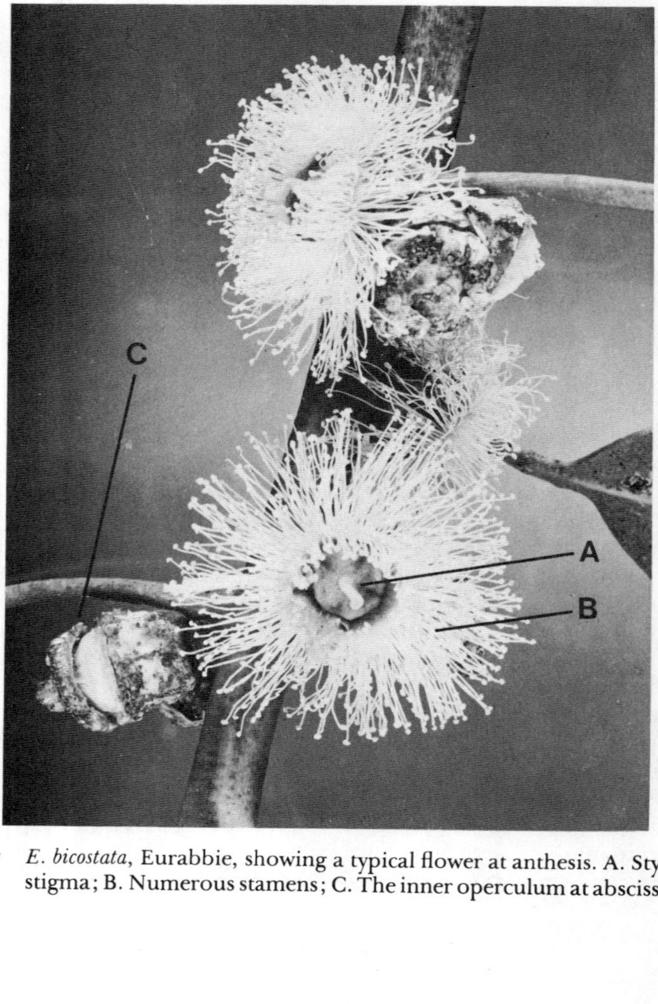

Fig. 2.7 *E. bicostata*, Eurabbie, showing a typical flower at anthesis. A. Style and stigma; B. Numerous stamens; C. The inner operculum at abscission.

characteristics such as the large number of stamens, and a common feature of the family in having a perigynous arrangement of the outer floral whorls (Fig. 2–7). The fruit which is produced after flowering is in most cases a quite hard woody capsule which in due course dries and dehisces shedding seed. In framing classifications, *Eucalyptus* was dealt with particularly by George Bentham in *Flora Australiensis* (BENTHAM, 1867) and Bentham in turn relied on specimens and information provided to a large extent by Ferdinand Mueller from Australia. In the herbarium, floral features have always been of particular value in classification and Bentham recognized differences in anther shape as well as in the arrangement of attachment of filament and anther. He used these characters in developing his classification of the genus. The generalized anther shape in *Eucalyptus* is that which is general for many flowering plants; a pair of anther 'cells' placed closely together with a narrow connective and dehiscing by longitudinal slits. A specialized type is with dehiscence by terminal pores in which the anther is adnate rather than versatile (Fig. 2–8a, b, c). Features such as these are useful in framing keys (BLAKELY, 1965) in classifications for grouping species or making a determination. The ovary characteristically has a number of locules, often four or five, but between three and about ten can be found in various species within which the ovules are located before fertilization. The ovules differ in their basic type being anatropous in one group but hemiptropous in the majority. Moreover, there are non-functional ovules, which have been termed ovulodes which, unlike the true ovules, never form seed. After the fertilization of the ovules, the ovulodes remain as small structures forming chaff. The seeds are sometimes morphologically very distinct from the chaff and at other times not readily distinguishable, although they are generally bigger and heavier. The number of viable, fully developed seed in any individual fruit is often quite small under natural conditions, four to ten per capsule being quite common but in some circumstances this may rise to twenty or thirty or even more. The seed is mostly quite small, 600 000–700 000 per kilogram being usual even when mixed with chaff and the amount of seed to chaff by weight is often as little as 5%. The seed when ripe will ordinarily germinate if exposed to suitable conditions of temperature and moisture within a few days, a week being common. In some species, especially those from the colder south-eastern mountains dormancy may develop, which is broken readily only by stratification, that is, treatment after moistening at temperatures of about 1°–2°C for a period of 4–6 weeks. In those species confined to areas where there is regular snowfall in south-eastern Australia and Tasmania, many respond to such stratification treatment which presumably simulates the field condition characteristic of an environment in which snow lies on the ground for some weeks each year.

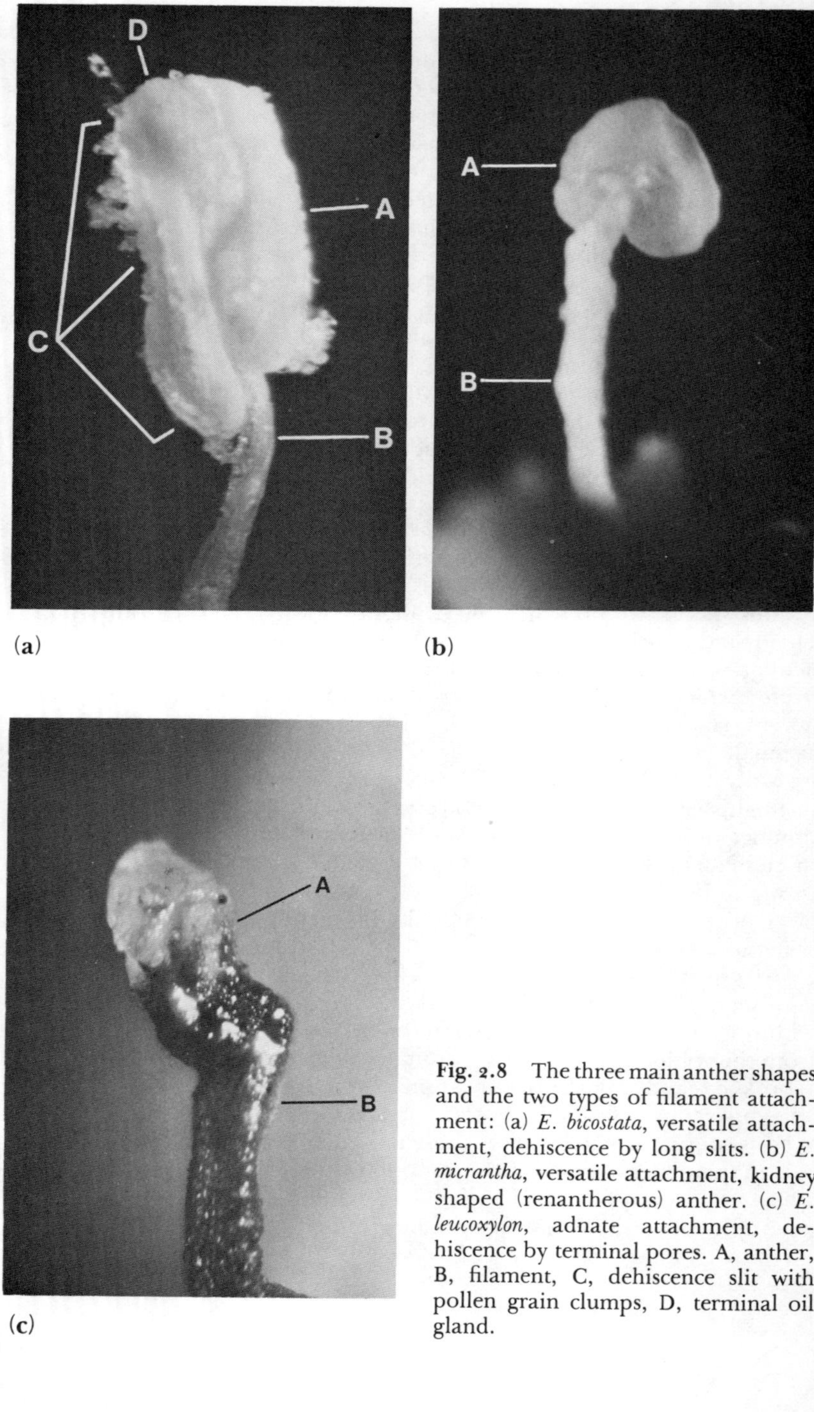

Fig. 2.8 The three main anther shapes and the two types of filament attachment: (a) *E. bicostata*, versatile attachment, dehiscence by long slits. (b) *E. micrantha*, versatile attachment, kidney shaped (renantherous) anther. (c) *E. leucoxylon*, adnate attachment, dehiscence by terminal pores. A, anther, B, filament, C, dehiscence slit with pollen grain clumps, D, terminal oil gland.

2.9 The inflorescence

The individual flowers of *Eucalyptus* occur as solitary units in the axil of a leaf in but very few species; *E. stoatei*, *vernicosa* and *globulus* being three cases. Mostly they are seen as a cluster of flowers in an inflorescence in the axil of a leaf in regular numbers of 3, 7 or 15 or more (PRYOR, 1954; MOGGI, 1963). From minute morphological evidence it is apparent that the *Eucalyptus* inflorescence is a cyme, very considerably reduced, but still in most cases with the branching structure characteristic of the regular cymose dichasium. The geometry of inflorescence arrangement in *Eucalyptus* is quite precise and fascinating. The inflorescence number is fairly consistent, especially in the lower number range. When the inflorescence is first discernible it is enveloped by bracts which completely enclose it. With growth and expansion of the developing flower cluster, the bracts are shed and the separate buds then appear. Depending on the particular subgeneric group, the time between the very first appearance of the complete inflorescence in the leaf axil and the shedding of the bracts may be a matter of a month or a year or more. The maximum number of individual flowers in an individual inflorescence may reach 30 or 40 in some species but the large majority are seen as in clusters of 3, 7 or 11 and 15. While each unit inflorescence is characteristically located in a leaf axil in some groups of species there is developed a compound inflorescence made up of these unit inflorescences in a particular manner more or less unique to *Eucalyptus*. The stem leaves are reduced or even almost completely suppressed, but the associated node persists in the alternate arrangement. These decussately placed leaf axils with their intranodal separation are the site of attachment of the peduncle of each unit inflorescence which are then aggregated into compound groups by shorting of the internodes and intranodes so that they form a complex cluster which resembles the classical panicle (Fig. 2–9). But this compound inflorescence is basically, as a branching system, like the leaf arrangement, that is a modification of the decussate branching pattern—and although known in the traditional descriptions as a paniculate inflorescence (or if flattened on the top, a corymbose inflorescence), it would perhaps be better termed a 'eucalyptoid compound inflorescence'. Those species which have a compound inflorescence tend to fall into discrete groups and this feature is useful for morphological characterization and diagnosis.

There is much variation in *Eucalyptus* in the size and shape of buds and fruits which, though relatively very constant in a given species, are substantially different between many pairs of species. Because these organs can usually be found on any specimen and because of their diversity according to species, much consideration has been given to them in systems of classification. The range of bud types is diverse and fruit

Fig. 2.9 The axillary and terminal inflorescence types. Left, *E. polyanthemos* (terminal); *E. elata* (axiliary).

characters vary widely as is seen partly in Figs. 5–1, 5–2. Although the *Eucalyptus* flower has no calyx or corolla at the time of anthesis, stamens are often prominent and showy, sometimes coloured and at times brilliantly so. Below the staminal ring at the top of the ovary near the base of the style there are nectaries which produce a copious flow of honey. Most eucalypts are undoubtedly insect pollinated and others are regularly visited by birds, while wind pollination is known certainly in one or two species and especially in *E. tereticornis*. The large amount of nectar produced has led to a substantial honey producing and export industry in Australia.

3 Classification

3.1 General

A main aim of classification in biology is to place species together in groups arranged in a hierarchy such that the similarity or likeness of any one species to another (or lack of it) is expressed by the position of each of them in the scheme.

By tradition, and from ordinary practical considerations, species are placed together in small groups when they share a larger number of morphological features in common than species which are placed in separate groups.

The number and names of groups applied to the classification of species within plant genera differ according to the features of the genus itself as well as to the inclinations of the authors, since there are no precise international rules which must be applied between the level of species and genus.

The latest *Eucalyptus* classification (PRYOR and JOHNSON, 1971) uses the groups (taxa) below the genus as follows: Subgenus, Section, Series, Subseries, Superspecies, Species and Subspecies. In the same scheme the seven species usually placed in the separate genus *Angophora* are considered as being so 'eucalyptoid' that they might well be included in the genus *Eucalyptus*. Proposals, such as this latter one, emphasize the strong anthropocentric influence in schemes of classification. The urge to group, classify and name organisms to help organize knowledge about them and to facilitate discussion has been a feature of human intellectual endeavour since such thought has been recorded, and perhaps before. It is easier to think of classification and the taxa involved and to appreciate both its value and limitations if this is borne in mind. There is no set of absolute values in this field (JOHNSON, 1970) and no very precise definition of such a basic taxon as 'species' has yet emerged, nor is it likely to do so. The concrete reality is the interbreeding population and the extent to which one considers such populations to belong to a single subspecies or species involves always a subjective element. Thus, the definition of species as 'that which a competent taxonomist considers a species', is not only a jest but it has some aspect of truth in it.

Tradition too, often plays a part in classification, so that some groupings tend to be retained because they were first proposed (MAIDEN, 1903–31) and no pressing reasons have since been found for abandoning them, even though there may be known alternatives which seem almost equally good.

The first thorough classification of *Eucalyptus* was by George Bentham

(BENTHAM, 1867) working in the Royal Botanic Gardens, Kew, with Australian herbarium material, much of which had been transmitted by Ferdinand von Mueller, who was Government Botanist of the State of Victoria. Bentham's classification is embodied in his treatment of *Eucalyptus* in his magnificent seven volume work, *Flora Australiensis*. Bentham gave a lot of emphasis to the morphology of the anther—a feature readily available to him from herbarium specimens, and this feature, though now being given less weight than he chose to give it still, to most workers, has a bearing on circumscribing some groupings.

At the other end of the scale, the classifiers have been the field botanists and countrymen who needed to give names to what they saw in the field and often wanted to use in building, for fencing or as fuel. The most prominent and distinctive characters to him were those of the bark and wood, so that the common names of the major subgeneric groups are derived from such field classification and are reflected in names such as Stringybark, Ironbark, Bloodwood and Ash.

3.2 Groupings in Eucalyptus

The most recent classification into subgenera and sections is set out in Table 1 which gives the common names too for the groups where they exist. The most important of these for preliminary study of the genus are the subgenera *Corymbia* (Bloodwoods), *Monocalyptus* (Stringybarks, Ashes, Peppermints) and in the large subgenus *Symphyomyrtus* the sections *Transversaria*, *Bisectaria*, *Exsertaria* (Red Gums), *Maidenaria* (Blue Gums) and *Adnataria* (Boxes and Ironbarks).

Some of the smaller groups shown in the table, but not included in the above list, have special interest because of the features they present which stimulate evolutionary speculation and suggest possible alternative ways of framing a classification. Authors such as Carr and Carr (CARR and CARR, 1959) would prefer to make a somewhat different arrangement at the subgeneric and sectional level, especially in relation to *Monocalyptus* as here circumscribed and some of the smaller subgenera such as *Eudesmia*, *Idiogenes* (PRYOR, JOHNSON, WHITECROSS, MCGILLIVRAY, 1967) and perhaps *Gaubaea*.

Once a classification has been proposed it is possible not only to scrutinize it using the same criteria at the same level as the authors themselves and thus to make judgments about its merits, but also to add additional criteria which from time to time become available from technical advances in biological and other techniques.

3.3 Aids to classification

All classifications propounded up to the present use macro- and micro-morphology to a great extent as has been the tradition back to the

Table 1 Principal subgeneric taxa in Genus *Eucalyptus*

SUBGENUS	*BLAKELLA*	*CORYMBIA*	*MONOCALYPTUS*	*SYMPHYOMYRTUS**						
SECTION				*Transversaria*	*Bisectaria*	*Dumaria*	*Exsertaria*	*Maidenaria*	*Adnataria*	*Sebaria*
VERNACULAR NAMES	Ghost Gums	Bloodwoods	Stringybarks Peppermints Ashes	Flooded Gum Red Mahogany Grey Gums	Mallees	Mallees	Red Gums	Gums	Boxes Ironbarks	Tallow Wood

* *SYMPHYOMYRTUS* is by far the largest Subgenus and the principal SECTIONS within it are set out.

time of Linnaeus and before, but in the more recent ones additional criteria have been employed with some success. For example in the classification by Blakely published in 1934 the species, *E. megacarpa* from Western Australia was placed in what is now called Section *Maidenaria* of the Subgenus *Symphyomyrtus*. This was particularly suggested to him because of anther morphology, a character to which Blakely, following Bentham gave great weight. Progeny assay from careful field collection disclosed, however, that *E. megacarpa* hybridizes naturally at times with *E. marginata* which is in the subgenus *Monocalyptus*. It was also found with anatomical study that the ovules and seeds (GAUBA and PRYOR, 1958) of *E. megacarpa* are like those of *Monocalyptus* and not of *Symphyomyrtus*. There is now general agreement that *E. megacarpa* has more affinity with species of *Monocalyptus* than with those of *Symphyomyrtus* and since *E. megacarpa* is closely related to *E. preissiana* and one or two other Western Australian species, placing of these in *Monocalyptus* is indicated by removing them from the Section *Maidenaria* of *Symphyomyrtus* where Blakely had placed them (PRYOR and JOHNSON, 1962).

Two points emerge from such study. First, the use of genetic and anatomical criteria may refine a classification which has been based only on morphology. Secondly, the weight to be given reasonably to a single morphological character, in this case the anther morphology, should be downgraded in certain contexts.

Another example is given by the phytochemical studies of Hillis (HILLIS, 1966a) based on an examination of phenolic extracts. Chemical substances of this kind are presumably the end products of various metabolic processes during the ontogeny of the plant. They are often produced very consistently in any one species. If the particular substance is not produced by other species, as sometimes happens, it may be used as a marker for grouping the species in which it occurs. Sections or even Subgenera may all possess a particular compound absent from other groups.

Hillis found consistently a substance in most of the species of *Monocalyptus* he examined and since this group was more or less equal to a group used in earlier classifications like Blakely's, which he called *Renanthereae* and as the substance was one not hitherto known Hillis called it renantherin. The presence of renantherin in *E. megacarpa* supported the proposed rearrangement based on morphological grounds.

There were reasons on other grounds for transferring a few species such as *E. deglupta* and *E. gamophylla*, which Blakely included in *Monocalyptus*, to *Symphyomyrtus* and *Eudesmia* respectively and Hillis in examining these species chemically, found no renantherin present in them. Phytochemistry thus supported the idea which had been propounded at about the same time, as a result of certain micro-morphological study that the species referred to had greater affinity with species of the Subgenus *Symphyomyrtus* than with *Monocalyptus*, and thus an example of the way in which phytochemical information bears on classification was set out.

Quite recently too, CARR and CARR, using the scanning electron-microscope were able to obtain precise pictures of minor sculpturings and excrescences on the leaf surfaces of eucalypts. These have not been seen so clearly before. They called these patterns 'phytoglyphs'. Phytoglyphs were found in some aspects to be consistent for one species and different but consistent in other species. By projecting information of this kind to a critical group—that is to say, a group of specimens which could not be readily classified—they were able in the species from tropical northern Australia ordinarily called *E. dichromophloia* to recognize four distinct taxa which might be designated subspecies or even separate species according to the judgment of the taxonomist considering the evidence. The essential point is that the precise form of the surface sculpturing was found, once it could be seen sufficiently clearly, to be an effective taxonomic aid in the material studied. Thus new methods or new equipment may have direct application by assisting in making classificatory judgments.

It is generally considered now that several other approaches of this kind will be rewardingly applied in the future to *Eucalyptus* classification. There is promise, for example, from such fields as biotic relationships as between insect parasite and host eucalypt, for eucalypts—like many other plants—have a large suite of insect species in the natural environment endemic to Australia and specifically attached to them. In some cases, as with the psyllids which are aphid-like Australian insects prevalent on *Eucalyptus*, each eucalypt species has more or less its own psyllid species (MOORE, 1970). The 'more or less' qualification is relevant, for it is questionable if a different psyllid is found on what seem to be plants of the same species, whether or not this can clearly indicate a taxonomic difference which has been overlooked on morphological grounds. In reverse too, if the same psyllid is on more than one species or subspecies, does this not suggest a critical scrutiny to see if the separation of the two plant taxa is really merited? The matter is further complicated because not all psyllids are host specific even within the above context. Some species of psyllid are more catholic in their tastes and may occur in several species of the same taxonomic group—such as subseries or superspecies.

Other approaches are indicated by recent developments in isoenzyme studies, which result from refinements in chemical techniques. A further example has been with the use of gas chromatography in the assay of essential oils in eucalypts (MCKERN, 1965).

3.4 Classification and phylogeny

Classifications have other uses. One to which they are often put is as a base from which to make phylogenetic speculations. Most workers prefer to erect a classification based on similarity or dissimilarity of taxa based on as many characters as can be observed with such weighting of the different characters as may seem desirable. Notions of phylogenetic lines of descent can be developed from the resulting arrangement, but it is dangerous to assume that a phylogenetic pattern can be seen in advance on which to construct a classification. The possibility of circular argument is always present if such an approach is adopted. In *Eucalyptus* some classifications in the past have put all species in a single 'tree' presuming descent from a single prototype, that is a monophyletic condition, but it is far from clear that this is really the case in the eucalypts.

From the generally accepted recent classifications some possible phylogenetic trends can be inferred, but there is still insufficient information to suggest at all convincingly that *Eucalyptus* is on the one hand monophyletic or alternatively polyphyletic. *Eucalyptus* (and *Angophora*) are distinctive plants and perhaps those features they display which mark them off from other plant groups tend to obscure a diverse phylogenetic origin, if they are indeed polyphyletic. To put it another way, no competent opinion has yet been put forward surely as to whether all that we now call *Eucalyptus* had a single common prototype ancestor, or whether more than one ancestral type relatively unrelated has given rise to descendants, which by developing in a similar direction, have produced species which resemble one another in spite of separated origins; that is to say, there has been evolutionary convergence. There is no significant fossil record to assist with such speculation and at present we note the question without providing an answer.

Classification has still other aspects. If species are related in the way that a 'natural' classification implies then the likelihood of any pair of species being able to interbreed will be related to the distance apart they stand in the system of classification. This has considerable bearing on matters of identification, breeding and ecological behaviour. Capacity to interbreed may also be used in the reverse way in framing a classification. In eucalypts this has been used in part but since the possibilities of interspecific crossings have been tested between relatively few pairs of potential combinations this aspect has not yet featured prominently in classification construction.

4 Breeding System and Manipulated Breeding

4.1 Reproductive processes

The *Eucalyptus* flower in its essential form is quite highly constant through all the species. At anthesis, the operculum abscinds and falls away. In the bioperculate species the outer operculum may be shed at an early stage of bud development or it may fall some time later, even at a time just before anthesis, to be followed shortly by the inner operculum. As soon as the inner operculum is shed the stamens which are either folded or appressed in bud, expand radially and the anthers dehisce exposing pollen (in almost all species) in irregular granular aggregates on the surface of the anther. The style is then surrounded by a disc of many stamens attached to the staminal ring. At this stage the stigma is usually still unexpanded and not receptive, although the pollen is mature and able to germinate if placed on a receptive stigmatic surface (Fig. 4–1). Thus the eucalypts are in general protandrous. The stigma usually becomes receptive one or more days after the stamens are fully expanded, by which time much of the pollen is already removed from the anthers by visiting insects.

Thus in any one flower the protandrous behaviour favours pollination from another source. However, this may still be from the same individual tree. Although in most cases any individual flower may endure some two or three days from opercular fall till stigmatic shrivelling, the time from the opening of the first flower on a single individual until the last may extend over one or two weeks or even longer. If the first and last flowering individuals are noted in a small interbreeding population of the same species the overall period may extend over four or five weeks or more. Thus in a single population the protandrous condition which in any case favours out-crossing is reinforced by the population behaviour in having some spread in peak of anthesis in different individuals.

As the stamens expand they carry the pollen in the anthers outwards and there is initially no transfer to the as yet unreceptive stigma. Pollination is in almost all species effected by vectors most of which are insects. In relation to insect pollination eucalypt flowers are not much specialized. This is reflected in the wide range of insects observed visiting them and presumably securing pollination. Various species of beetle, flies, both syrphids and calliphorids, bees (and nowadays introduced), native as well as other kinds are seen. In other cases, especially in those species with larger flowers, birds such as honeyeaters and brush-tongued parrots are likely to effect pollination. There is, however, only one well-

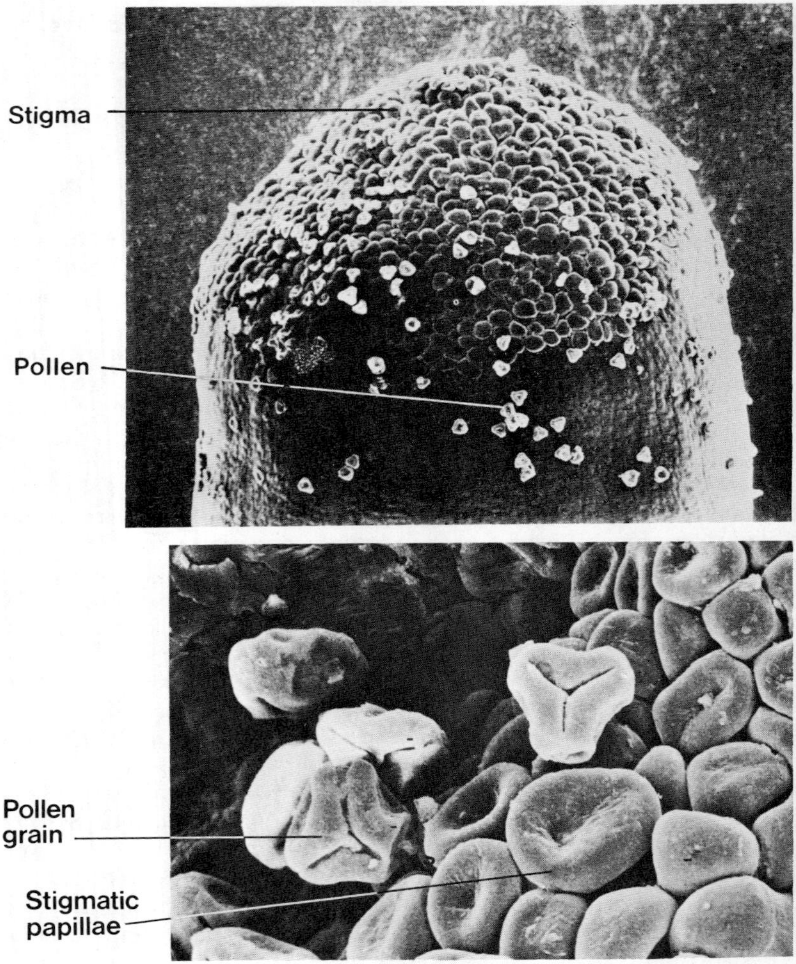

Fig. 4.1 Pollen grains and stigma of *Eucalyptus*.

documented case where wind pollination is effective and that is *E. tereticornis* which was observed to behave this way from the studies of COOLING and ENDEAN, 1966, on trees of the species planted in Zambia. The lack of adherence of the pollen in clumps in *E. tereticornis* which is so characteristic of most other eucalypt species is no doubt associated with this behaviour. But the presence of nectar and the retention of the general floral structure adapted to the entomophilous pollination may allow the species to retain partial insect pollination as well as wind pollination. There has been insufficient examination of eucalypts to say whether *E. tereticornis* is unique in this way, but from the appearance of pollen in a related species such as *E. blakelyi* it may behave similarly.

4.2 Incompatibility

In many flowering plants the likelihood of out-crossing is reinforced by a self-incompatibility mechanism. In the few eucalypts examined critically a system of this kind, determined by inheritance, has been found. In *E. pulverulenta*, for example, a 10 × 10 diallel cross shows that two individuals are self-compatible and will regularly set seed when selfed, that is, when pollinated with their own pollen, whereas three or four others cannot be induced to set seed at all by selfing and the remainder may set a little seed at times with difficulty if the only source of pollen reaching the stigma is its own; that is, it can be described as weakly self-compatible. In such individuals if its own pollen and that from another individual of the same species are present at the same time on the stigma the seed set is likely to be mainly by the pollen from the external source.

Such a breeding system may be described as one of preferential out-crossing which is reinforced by a gene-controlled incompatibility mechanism which impedes or prevents selfing and this is also in its operation aided by the protandrous condition in which the anthers dehisce and pollen is shed from each flower before its stigma is receptive. The evidence from trees isolated to the extent that seed set can result only from selfing (usually ornamental planted trees in isolated locations) suggests that the extent of incompatibility differs somewhat in different species and different systematic groups of species. But in principle it seems to be general through the genus. In one or two species, it has been suggested that there may exist an inbreeding system although none has positively been identified and if such does occur in any species it must be rare in the genus.

4.3 Mean free pollen path

The existence of an outbreeding system implies that considerable genetic diversity is retained in interbreeding populations. The size of an interbreeding population will be affected by the mean free pollen path which, of course, is dependent on the vector involved.

This parameter is best estimated by the use of a genetic marker. That is, by identifying a readily recognizable inherited feature displayed preferably in young seedlings by inheritance as a dominant trait. This has been examined most precisely in assessing the extent of interspecific crossing since F_1 hybrid seedlings are usually readily distinguished from non-hybrid parental-type seedlings. An example is provided by *E. fastigata* and *E. robertsonii* which in most years do not reach anthesis at the same date but do occasionally flower coincidentally. By collecting seed from parents which have flowered concurrently and taking pairs of species, one of either parent separated by varying distances, some estimate of the pollen dispersal is given by the number of interspecific hybrids in the progeny. Such hybrids are readily recognized in the above combination because the juvenile leaves are radically different—in *E. robertsonii* at node 8 they are opposite and sessile while in *E. fastigata* they are alternate and petiolate. By scoring them in the progeny raised from trees located at different spacings the effective distance of pollen travel is indicated in Fig. 4–2.

A somewhat similar figure may be inferred from Cooling's studies (COOLING and ENDEAN, 1966) of *E. tereticornis* even though this species is in some degree, if not totally, wind pollinated.

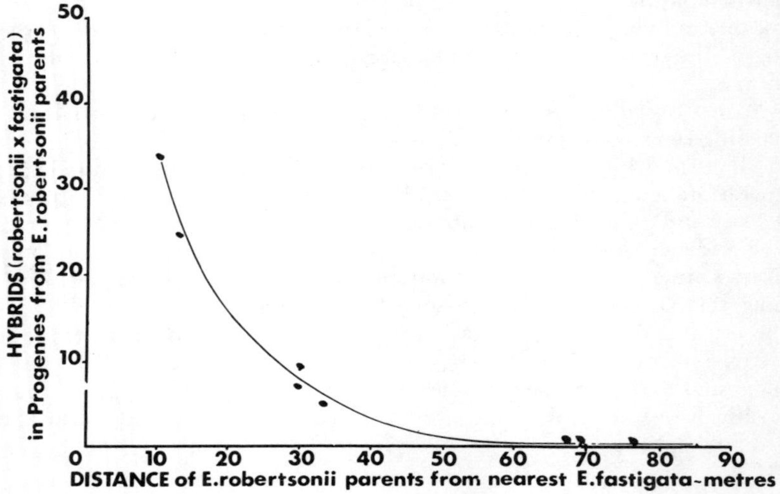

Fig. 4·2 The curve illustrates the distance of pollen travel between pairs of trees as revealed by hydrid progeny in *E. robertsonii* seed lots contaminated by pollen of *E. fastigata*.

4·4 Pollen behaviour

Eucalypt pollen compared with that of some flowering plants, e.g. grasses or vines, is robust and retains viability under ordinary ambient conditions if it remains reasonably dry, for some days or even weeks after anther dehiscence. The best way to store it for longer periods is to desiccate it over silica gel for 24–48 hours, either together with the anther tissue or free from it if it can be separated readily. If put in a sealed tube when dry and placed in a deep freeze it will retain capacity to germinate for more than a year, and if kept at −50° it has been found to be still functional after three years and perhaps will remain so for much longer. Viability can be assessed readily by placing a pollen sample on an agar plate with a suitable sucrose content and incubating at 23°C. Germination is seen at six hours and generally will be complete in 24 hours.

In vivo, germination on the stigma is relatively rapid and in compatible crosses the pollen tube grows through the style to reach an ovule and effect fertilization. The ovary contains many ovules but relatively few ordinarily set seed. Usually about 3–8 seeds per capsule are produced which is not a very efficient reproductive rate as is evidenced by the results of manipulated pollinations with the same species when as many as 20–30 seeds may be produced. When such high numbers of seeds are set many of them may be small and will produce seedlings with low vigour of growth.

4.5 Inbreeding depression

While selfing can be observed, often enough the results are seedlings of less vigour, and very marked depression of growth following inbreeding has been observed in some cases. Thus while as a partial survival mechanism, self-compatibility may make a useful contribution, it is clear that the weaker growth of selfings would make them of value in terms of biological survival only if the more vigorous out-crossed individuals had for some reason failed to appear in the regenerating population. In a normal situation of random crossing or preferential out-crossing any selfings which might by chance be produced occasionally would be soon lost in the consequent seedling population because of their lack of vigour.

An indication as to whether pollen is viable or not can also be gained by an immunofluorescence technique to determine whether the proteins present in the pollen grain wall when the pollen is normal and viable are still in situ. If subjected to germinating conditions, there is in a short time a release of proteins from the wall of the grain. If the germinating medium contains antigenic serum produced by injecting a rabbit with the same pollen the protein is soon immobilized on the plate and by attaching to it a flourescent material the presence of the protein is readily detected by exposing the plate to U.V. light which gives a strong fluorescence. Occasionally, in some individuals of some species, as in *E. grandis*, no protein is released from the pollen grain wall and the grains also do not germinate on sucrose agar. Plants like this are effectively male sterile but the female component functions normally with viable pollen from another plant and will set seed in the usual way. In effect then the flowers on a single plant which produces abnormal and sterile pollen are biologically unisexual.

Another form of pollen sterility has been found in *E. pulverulenta*. In some individuals the pollen produced is defective in that the grains when shed have no protoplasmic contents. The condition is determined by inheritance and has been found at a low frequency in a natural population which has been studied.

If viable pollen is placed on a stigma with which it is incompatible, fertilization does not take place and it appears that the precise reactions resemble those which have been studied closely in other plant genera, although the details have not yet been fully explored in *Eucalyptus*.

5 Interspecific Hybridization

5.1 General

Hybridizing between pairs of species is fairly common in plants and often they are fertile, producing viable seed and seedlings (STEBBINS, 1950). If such behaviour were unrestrained the consequent genetic mixing would lead both to loss of species identity and pangenic populations, which might be regarded as a single species. Since such extensive genetic mixing does not happen, even though some interbreeding can and does occur, it is evident that factors operate to ensure the maintenance of species identity and reduce or inhibit complete genetic mixing which would be part of such a condition. The point at which inhibition of interbreeding or of the successful performance of resulting hybrid progeny occurs differs in separate situations.

Many pairs of species of Eucalyptus can and do hybridize (BLAKE, 1953; JACKSON, 1958), but except in some limited situations which often, but not always, are a result of the impact of European settlement and land use in Australia, the species do not break down genetically and the products of interspecific breeding remain few and scattered relative to the total species population numbers (BRETT, 1938).

5.2 Form and occurrence of interspecific hybrids

To study hybridization in natural populations one must learn to recognize hybrids.

The location and appearance of interspecific hybrids is associated with some regular patterns or rules which have been put together after the examination of many actual examples.

Since identification of species is ordinarily by their macroscopic morphology, the first clue that an individual might be a hybrid is usually the recognition of morphological features which are intermediate in some degree between two species which might be supposed to be its parents.

Most morphological features are subject to multiple gene control so that if for example, in two species there is one with large fruits and the other with small ones the first generation hybrid, the F_1, will have fruits which are in size somewhere between those of the two parents. This is well illustrated by the hybrid *Eucalyptus viminalis* (small fruit, say 8 mm diameter) and *E. globulus* (large fruit about 20 mm diameter). The F_1 hybrid has fruit about 12 mm diameter. It is common in such size relationships for the hybrid to be in a position of the logarithmic mean rather than the arithmetic mean so that in the above case 12 mm is the hybrid dimension rather than 14 mm. If one can examine an F_2 generation it will be found that there is a complete range in fruit size from almost one parental type to the other which is a consequence of multiple gene inheritance.

There are numerous examples of such size intermediacy in other organs in hybrids, such as the length of the leaf petiole, leaf size and pedicel or peduncle length in the inflorescence (Figs. 5–1, 5–2).

In qualitative characteristics too a similar condition exists. To use the same example, the flower buds of *E. globulus* are highly glaucous when they first appear from the enclosing bracts due to a covering of wax in rods or granules of particular physical character on the cuticular surface of the operculum, while those of *E. viminalis* are different and appear non-glaucous or green. The F_1 hybrid is somewhat but not highly glaucous and this at a first approximation would be described as of intermediate glaucousness relative to the two parents.

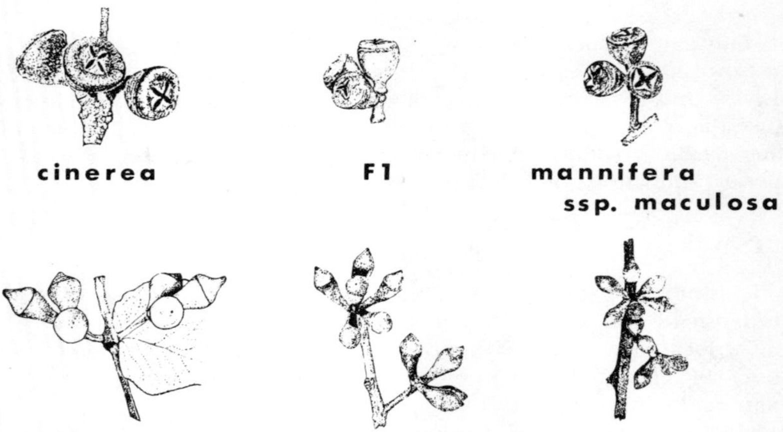

Fig. 5.1 The shape and size of fruits (upper row) and buds (lower row) in the two parents *E. cinerea* (Argyle Apple) and *E. mannifera* ssp. *maculosa* (White Gum) in relation to their F_1 hybrid which was produced artificially.

A similar condition which is easily recognized in the field is that in which two species of a different bark character are involved. F_1 hybrids occur between *E. rossii*, which is a gum with smooth decorticating bark, and *E. macrorhyncha* which is a stringybark and has a rough persistent furrowed fibrous outerbark extending throughout the length of the trunk and branches almost to their leafy extremities. The hybrid is said to be 'half-barked' and is of intermediate character. It has a rough stringybark about half or two-thirds of the bole height but the upper bole and branches are gum-barked and decorticate regularly. Where species which have such contrasting characters are involved in hybridizing, the field search for hybrid off-spring is much simpler than when the parents are similar in field diagnostic morphological characters, so that the recognition in the field as a first step in locating a hybrid such as between *E. viminalis* and *E. dalrympleana* will be difficult unless some juvenile or intermediate leaves can be found, since this is a character in which the two species differ markedly. Their general appearance is similar in that they are tall, white gums with a very similar field appearance. It is more difficult still to be sure that one has a hybrid between *E. dalrympleana* and

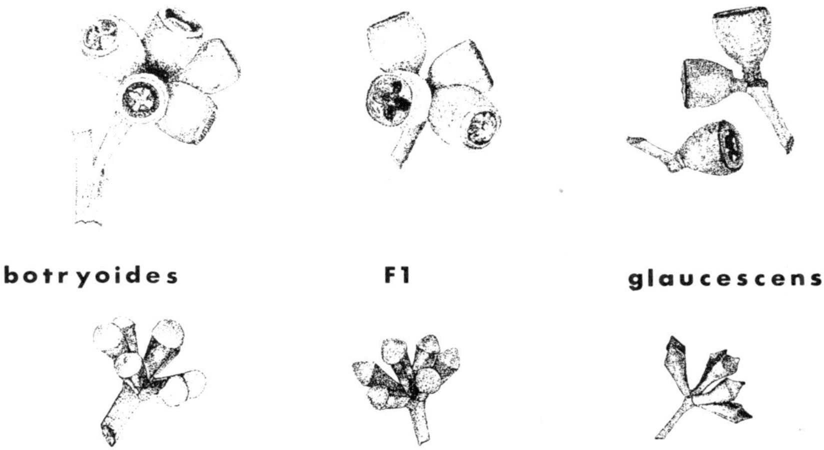

Fig. 5.2 As for Fig. 4 except the species involved are *E. botryoides* (Bangalay) and *E. glaucescens*.

E. rubida which, though distinct species, are similar in the shape of their juvenile leaves too, as well as in bark character and number of buds in the inflorescence.

Thus a first requirement in determining a specimen as hybrid is to observe a display of intermediate morphology. This may be refined and taken to an anatomical level if warranted, or examined statistically. It may also be supported by other characters, such as the nature of the chemical constituents of the plant. This latter approach in looking for intermediate characters can be made chemically precise but is sometimes indicated in the field so far as one chemical aspect is concerned by an intermediacy in essential oil characteristics, which, with practice, can reasonably well be inferred by the smell of crushed leaves.

A second feature which can be taken to support the notion that there are hybrid individuals to be found, follows from the precise location in which the hybrid occurs. The supposition of hybridity will be more likely if both putative parents are also seen in the same place not too distant from each other to hinder cross pollination.

A further way to recognize hybrids is by an assay of the progeny of presumed interspecific intermediates. This can be done by manipulated crossings, but that is a long and rather tedious process often not warranted for the purpose, so that recourse to raising progency from seed from open pollination is the rule. Most eucalypt hybrids are fertile and usually a little ripe fruit can be collected and seed extracted from it. With this material the maternal parent is surely known and while the seed is possibly the result of self-pollination it is more likely to be the result of back-crossing to one or other of the supposed parents or perhaps a second similar hybrid if there is one nearby.

While there is a loss of precision compared with the examination of progeny from fully controlled pollination, the study of progeny from open pollination from a supposed hybrid can yield much information. Progeny testing is also especially valuable since the characters of the juvenile foliage are usually well displayed by the time the seedlings are four months old and in many hybrid combinations the juvenile leaf characters are strongly in contrast between the two parents. Whether the offspring are from selfing, back-crossing or are an F_2 generation, the display of some segregation and recombination of characters in the seedling population is not prevented, although the numerical proportions will be closely related to the genetic make up of the originally supposed hybrid and the unknown male (pollen) parent.

Contrasting juvenile characters are illustrated by considering the combination of *E. dives* and *E. pauciflora* between which species, hybrids are well known. *E. dives* has, amongst other things sessile, ovate-lanceolate, glaucous juvenile leaves while *E. pauciflora* has petiolate, elliptical non-glaucous ones. Each character segregates independently and is capable of recombination, although evidently not always in simple mendelian ratios because of the graded conditions resulting from multiple gene control of each such characters. Thus, if the population is large enough some seedlings will show ovate-lanceolate leaves of the *E. dives* shape without being glaucous but also with a petiole, while others will be elliptical and non-glaucous but sessile. The first two characters in this latter case being those of *E. pauciflora* while the latter is that of *E. dives*. If the progeny from a supposed hybrid displays such variation the probability that it is, in fact, derived by breeding between the putative parents is considerably greater than when the supposition has been made only on the basis of morphology and location. To make a still more rigorous assessment, if there is sufficient time and need, the supposed hybrid can be produced by manipulation between the parents and a replica of the naturally spontaneously occurring individual made as a result of manipulations. In a few instances this has been done, as for example, between *E. blakelyi* and *E. cinerea* where the spontaneously-occurring hybrid individual was matched by one raised from the two parents using *E. cinerea* as the female parent.

5·3 Breeding barriers

5.3.1 Geographic barriers

It was pointed out that while hybrids occur between some pairs of species this has not lead generally to a breakdown of species integrity in the natural Australian scene. There are various barriers which evidently inhibit genetic breakdown. The simplest is geographic isolation and it is known from direct experiment that species pairs, one of which occurs in Western Australia and the other in Eastern Australia, can in some instances hybridize and produce viable offspring, but it is clear that under natural circumstances they have never been exposed to one another and the hybrids have never occurred spontaneously. An example is *E. caesia* × *E. pulverulenta* which was produced experimentally to verify this point. The first species has limited occurrence in Western Australia while the second is equally limited in eastern New South Wales. Geographic isolation is also effective over shorter distances than the 2000 miles separating Eastern and Western Australia with the large gap in climatic and soil zones suited to pairs of species such as the above two. Apparently equally effective isolation is provided by the altitudinal separation of, for example, *E. delegatensis* which does not generally occur below about 1000 m and *E. pilularis* which, in latitude 35° South in south-east Australia is not found above 300 m elevation in eastern New South Wales in the same latitude. Again, a viable hybrid has been produced by manipulation between these two species, but there has evidently never been an opportunity under natural conditions for crossing to occur between them.

5.3.2 Flowering time

To hybridize, species must flower at the same time. In the date of anthesis there are variations within eucalypts. All species to some degree flower by season. Some reach anthesis in very close accord with date, suggesting that there is some daylength control of their flowering mechanism. Others vary from year to year and while, for example, they may flower in late spring, early summer or high summer, the precise time seems to depend upon the more variable features of the environment like temperature, soil moisture and possibly on something internal in the plant itself. Some species are exclusively autumn flowering in southern Australia, others exclusively spring flowering and though it is known as a result of manipulation involving the storage of pollen that certain pairs of these can produce viable seed and viable hybrid individuals, they do not ordinarily hybridize. A very good example of this is the pair of species *E. stellulata* and *E. pauciflora*. *E. pauciflora* is widespread in south-eastern Australia and has a wide altitudinal range. At lower altitudes it commences flowering in early spring, e.g. in southern Australia in September, and at higher altitudes concludes in mid-summer about the end of December. *E. stellulata* on the other hand is a strictly autumn-flowering species and seldom reaches anthesis in any part of its range before the middle of March and often it is during April. Thus these two species, by flowering time alone, are reproductively isolated even though through a wide part of their ranges they each occupy their own site in an intricate mosaic with very extensive common boundaries where individuals of the two species are closely juxtaposed and where pollen exchange could take place readily. The only natural hybrids recorded between these species have been from the Barrington Tops in New South Wales, a locality which is an isolated plateau in a geographic situation which represents marginal conditions for the range of both species. Whilst the examination has not been made, there is perhaps some change of flowering time in one or other of them that causes them to overlap more often than is the case in the more extensive part of their range, where it never occurs. By storing pollen of *E. stellulata* in a deep freeze over winter F_1 hybrids have been manufactured between the two species so that it is known they can hybridize successfully. There is therefore almost complete breeding isolation provided by flowering time differences, as is probably the case with other species pairs.

5.3.3 Systematic affinity

It is evident also that in general the subgeneric groups of species are probably completely reproductively isolated from one another for while by no means all of the possible combinations have been tested the available evidence points to the fact that many species pairs either in the field or by manipulation will not produce hybrids. This is the case in any of the three subgenera *Corymbia*, *Monocalyptus* or *Symphyomyrtus*. Within the large group *Symphyomyrtus* there is a high level of isolation between *Adnataria* and the other sections. A few interspecific hybrids have been found between species of *Adnataria* and a species in some other section of *Symphyomyrtus*, but although intermediate in morphology they have not produced viable seed or progeny. An example of this is in the combination *E. largiflorens* × *E. camaldulensis* of which, on morphological grounds, a few trees are known in the vicinity of Deniliquin, New South Wales. A few have also been produced by manipulation such as *E. bicostata* × *E. melliodora*.

Thus the relationship between species, according to the particular systematic grouping to which they belong, is an index of likely capacity to interbreed. Interspecific hybrids between the subgeneric groups have not yet been found or manipulated.

5.3.4 Ecological isolation

There are many species in the same subgenus which, while occupying their own particular habitat, occur in stands adjoining another species. The separate distributions apparently are determined generally by the habitat conditions. It is in such cases at the stand margins that hybrids are frequently found. In the case of *E. pauciflora* and *E. dives*, for example, a common distribution pattern in some areas follows geological substrate, *E. dives* occurring on small ridges formed by metamorphic sedimentary rocks of shale, slate or phyllite occurring as outliers in a granite landscape. *E. pauciflora* in these circumstances is confined to granite and the two species are often found side by side where their respective population boundaries meet. At this line of junction of their two populations it is usually easy enough with a little search to find an interspecific hybrid or two. In such situations it is quite easy to imagine that while the hybrid is vigorous and viable it does not oust either of the two parent species from their preferred habitats, since each on their own site seems to be capable of performing better than the hybrid. There is no graded intermediate situation which favours the hybrid combination in the geological context of this specific case, where the change from sedimentary to granite substrate occurs abruptly in a few feet with almost equally abrupt soil changes. Thus there is almost no intermediate habitat to favour the hybrid.

There are other circumstances as for example between *E. dalrympleana* and *E. viminalis* or *E. fastigata* and *E. delegatensis* where there are similar junctions between species populations without there being a substrate separation. In these two cases the habitat change is one graded with altitude and yet the stands form sharp boundaries. Again interspecific hybrids can occasionally be found at such junctions but they do not spread into the area occupied by the general species population. Such barriers to the survival of the products of interbreeding might be described as ecological barriers and it seems evident that in a great many cases in *Eucalyptus* this is the kind of barrier which has prevented a genetic breakdown between discrete species. There are also hints of other barriers but they have not been studied sufficiently to describe them precisely. For example, *E. robertsonii* and *E. dives* are closely related species which sometimes flower at the same time and it is known they can interbreed, but hybrids between them have not been found over most of the field range through which these two species occur on their appropriate sites, but where they form very extensive margins one with the other in a complete mosaic. Some interbreeding barrier is evidently operating to keep them apart.

5·4 The fate of products of hybridization

If there is no significant survival of progeny derived from hybrids the hybrid individuals live out their life and finally die. However, there seems to be a low frequency of successful production of F_1 hybrids so that one finds a more or less constant low percentage of them at the appropriate sites as described above. Hybrids between species are more common in southern Australia than northern Australia and possibly this reflects the fact that there has been more change following European settlement in the south than in the tropical areas of the north, but there is also less opportunity for the formation of hybrids in the northern third of the country because there are fewer species there anyway, although the general pattern and opportunities are of the same kind. If there is production of progeny from established hybrids, presumably frequently from back-crossing to one or other of the supposed parents and the ecological barrier is less effective than is often the case, some spread of hybrid progeny as recombinants following back-crossing may occur and be fairly extensive, leading to the survival of whole populations of hybrid individuals producing a hybrid swarm as has been described for other plant genera in different parts of the world (ANDERSON, 1953). It seems evident that the process of spread of the products of hybridization is mostly associated with ecological disturbance of the natural plant communities, mainly due to the direct or indirect effect of European settlement either by partial land clearing or the use of fire. It appears the ecological limitation to the spread of hybrids in most cases is only reduced if there has been some form of destructive activity which limits its effectiveness. However, it is thought that there may have been some natural events which at times have produced a similar effect associated particularly with changes in the Australian climate, especially as reflected in a geographic shift in weather patterns so that zones which at present lie within a particular rainfall regime may have been drier in the immediate past or wetter at a period before that. If these projections are true then the boundaries of species populations may have shifted and in some circumstances hybrid swarms may have been released from foregoing hybrid parents and the swarms may have been somewhat better adapted to occupy the now changed habitat which was previously fully occupied and held by the pure species. In a number of places it is suggested that tension zones of this sort exist and since such a variation pattern may be paralleled in two or three groups of species this is a plausible explanation to give if climatic change is assumed. It seems a special circumstance may arise at times as a result of such climatic changes which cause the withdrawal of a species from a particular area, leaving behind some hybrid remnants with a second parent now more resistant and able to thrive in the changed zone, so that the second species occurs only as displayed by its genetic remnant in populations which are otherwise the alternate parent. Genetic remnants of this sort have been called 'phantoms' and are found at times at distances up to 50 km or even 100 km from the nearest pure population of the missing parent species. It is thought that such distances are too great for the situation to have arisen as the result of a chance long range pollination, even if birds were involved.

Subjects like this are rather speculative and perhaps will be explained more certainly in future investigations.

5·5 The significance of interspecific hybrids

The recognition of hybrids when they occur is necessary for precision in identification, both in the field and herbarium and also with eucalypts as exotics planted outside Australia.

Some binomials have been based on hybrid type specimens and these, since they purport to describe species, must be discarded because hybrids cannot be equated with species. Thus in the literature there are some fifty or so binomials which cannot be the names of valid species. For example, *E. vitrea*, *E. oxypoma*, *E. jugalis*, *E. laseronii*, *E. kirtoniana*, *E. nowraensis*, *E. kalangadooensis*, *E. kalganensis*, *E. edneana*, *E. congener* and many others are hybrid. Where the hybrid status is surely known the notation ×*E. vitrea* derived from the binomial can be used or the plants alternatively can be described by their parentage as in ×*E. nowraensis* which would be substituted by the description *E. gummifera* × *E. maculata*. The detection of hybrids and the application of this knowledge to identification may therefore be of considerable importance.

Where interspecific hybridizing is expressed as extensive hybrid swarms a position is recognized which is intractable in terms of ordinary taxonomy and must be handled by supplementary descriptions if an accurate statement about the populations is to be made. Hybrid swarms can provide information which may allow some development of ideas about the evolutionary history of the groups and of aspects of plant geography (STEBBINS, 1950). These will be dealt with later. Also when eucalypts are taken from their natural habitats and planted in collections either for scientific or commercial use in exotic habitats outside Australia the likelihood of hybridizing is increased for a number of reasons, so that matters of precise identification, genetic stability and cultivated use of them is complicated and frequently will only be satisfactory if basic knowledge of the hybridizing patterns and the results of it are known and taken into account.

6 Site Adaptation and Clinal Variation

6.1 General

The total habitat occupied by the various species taken together is diverse. Apart from the full geographic spread, described in Chapter 1, there is considerable altitudinal diversity—from sea level to 2000 m in south-eastern Australia together with a variety of substrate types giving soils ranging from light textured and acid to heavy textured and alkaline, which are occupied by *Eucalyptus*.

Physiographic differences, particularly those of aspect, lead to site differences which are reflected in the species distribution. Thus whole continuous areas of *Eucalyptus* vegetation are a mosaic of many species.

Some species have a very restricted occurrence as, for example, *E. ficifolia* in Western Australia which occurs in a few patches occupying no more than perhaps 25–50 square kilometres or *E. crenulata* which is known from three small sites in Victoria, covering a mere 40–80 ha. while *E. morrisbyi* occurs on a few hectares on the east coast of Tasmania and *E. pachycalyx* is found in an area of only a few square kilometres near Herberton in North Queensland.

These species with restricted distribution pose special problems in seeking an explanation as to their derivation. They frequently suggest a relict condition—that is survivors on special sites where they are able to evade competition from others and perhaps representing the remnants of much more widespread populations of some past time, when perhaps there were climatic conditions somewhat different from the present.

While these species of restricted distribution are of interest, they are the exceptional types; the common and more general situation is that each species has a fairly large geographic spread. The most widely distributed species are, *E. camaldulensis* and *E. tereticornis*. The former occurs in all mainland States and from the NW. coast of Western Australia to within 240 km of Darwin, to the foot of the dividing range in the eastern States, at times scarcely more than 160 km from the east coast, and to the south it reaches the coast in SW. Victoria and South Australia.

The latter species, *tereticornis* is confined to east coast regions but has a very wide latitudinal spread, being found in Papua at about latitude 8°, right through to SE. Victoria in about latitude 38°.

Distributional ranges of 1000 km or more are common and in some cases altitudinal spread is considerable, 1000 m being common.

It is evident that where a species population has considerable geographic spread, some of the habitat factors will differ in different parts of its range. For example, the difference between maximum and minimum day lengths for the year is much less in latitude 8° than in latitude 38°. The minimum temperatures experienced are much lower at say 1500 m than they are at 500 m and sites nearer the east coast generally have more rainfall than sites further inland, as for example the inland side of the Dividing Range.

In considering species which are widespread, one can often find morphological and physiological differences (LARSEN, 1965) which are correlated with a graded habitat factor.

It is useful to study the correlation of quantitative variation in morphological features with altitude differences. Altitude is, of course, correlated with rainfall and temperature variation as well as other habitat factors.

Where habitat factors are graded regularly, as with altitudinal change, and some morphological feature in a species is found to vary in a graded quantitative way which is correlated with that varying habitat factor, then the resulting condition is that (using Huxley's term (HUXLEY, 1940)) of a cline, or we can say that a particular feature in a species is showing clinal variation.

In eucalypts clinal variation in one or more features in relation to different habitat gradients is common and some of these have been studied in detail (BARBER and JACKSON, 1957).

6.2 Altitudinal clines

One of these studies is *E. pauciflora* (PRYOR, 1957) which occurs mostly as a pure stand of that single species on the mountain ranges of SE. Australia, especially in the altitudinal range from 1300 m to 2000 m. The species also occurs at other altitudes and in other locations but the variation pattern then becomes more complicated and more difficult to interpret.

From 1300 m to 2000 m there are regular changes in several morphological features. By careful sampling, continuous relationships between several characters and altitude are disclosed.

At 1300 m the trees are tall, with long leaves, small fruits, non-glaucous stems and thick bark, whereas at 2000 m these characters are expressed in the reverse way—short stature, short leaves, large fruit, glaucous stems and thin bark.

At 1650 m the trees are intermediate between the extremes for all these characters (Fig. 6–1).

The effect of altitude on temperature is the general one that there is a close direct correlation between mean monthly minima and maxima with altitude. Also over this altitudinal range total annual rainfall increases regularly with altitude.

Since there is no similar graded condition of the geological substrate and the soil resulting from it, it is inferred that the variation seen in the

Fig. 6.1 Clinal variation as illustrated by *E. pauciflora*. *Above left*, 1300 m; *above right*, 1650 m; *left*, 1900 m.

trees is the result of the adaptation of the plants to the habitat and that in some way the features displayed by the trees at the high altitudes make them better fitted to survive in that habitat than otherwise; the same situation obtains at low altitudes or any of the intermediate locations.

It is not always easy to establish in an experimental system just what survival advantage a particular feature confers on an individual. For example, it is not at once evident why thin bark should be an advantage at higher altitudes instead of thick bark, but on the other hand, one can suppose that shorter leaves may be able to withstand snow or ice loading without breaking as readily as longer ones and glaucousness in leaves or stems can be demonstrated as rendering those organs less likely to suffer injury in freezing conditions. In the physiological features experiment is more readily possible. There is seen to be from reciprocal transplant experiments at both high and low altitudes a good correlation between capacity to resist low temperatures and altitude of seed origin—the higher being the more cold resistant. In planting seedlings at 1700 m, only those from 1800 m and 1700 m and some from 1500 m survived, while those from 1300 m and 1200 m were killed out completely during a trial period of two years.

Likewise rate of growth in height is seen to be strongly inherited—those derived from seed from the higher altitudes being slower growing when planted at 1200 m than those from the lower altitudes. In the field, of course, this can be tested only at the lower altitude site since death from low temperature at the higher altitude sites eliminates the seedlings with a lower altitude origin.

Clinal variation in species which extend over 300 m of altitudinal range is general in both morphological and physiological features although the contrast between the extremes may not be as great as in the case described.

Similar gradients have been described in detail by BARBER and JACKSON in Tasmania and physiological variation in low temperature resistance has been tested experimentally in *E. fastigata*.

6.3 Clines of 'continentality'

A commonly found environmental gradient in Australia is in proceeding from the coast inland. The climatic features take on a more continental aspect the further the distance from the coast without any altitudinal change. Temperatures especially show wider diurnal and seasonal variation and rainfall generally diminishes.

Some species have a sufficiently wide geographic distribution to extend from coastal or near coastal localities to some hundreds of kilometres inland.

The clinal patterns which they display have been observed but not yet studied in detail.

At the inland limit species often carry thicker leaves, smaller organs such as fruits and are sometimes more glaucous.

6.4 Clines and taxonomy

One of the results from recognizing clinal variation is in taxonomic treatment and the case of *E. pauciflora* and *E. niphophila* is a good example.

E. niphophila was described from about 1700 m on the Mt. Kosciusko range. It is the form of *E. pauciflora* which is developed towards the upper altitudinal limits of the cline in this species.

There is, of course, no discontinuity in morphological variation between the form described as *E. niphophila* and *E. pauciflora* and it is not biologically acceptable to separate Snow Gums into two species on this basis. Thus the species would retain the name *E. pauciflora* and *E. niphophila* would become invalid as a species name. Since botanical nomenclature not only depends on biological facts but also on botanical law, this rearrangement would be adopted only after publication of the proposal and the decision of users to adopt it.

But even so, the matter cannot end here because there is a need to retain the information which is contained in the label *E. niphophila* and it is thus desirable that a means be found to describe the different parts of the clines where clinal variation is at a sufficient level to make this necessary for general use.

The name *E. niphophila* can be said to describe a reference point in the cline which can be called a 'Cline Form'.

The formal rules of botanical nomenclature do not yet adequately meet this situation and in the case of *E. niphophila* the application of them might result in reducing it to a 'variety' or 'subspecies' but this would not adequately describe the true situation.

For the present the position can perhaps be made clear, when this is necessary, with an explanatory note.

Some thirty binomials in *Eucalyptus* are regarded as cline forms, although they have not been reduced to that status by any formal publication in accordance with the International Rules of Botanical Nomenclature.

6.5 Site adaptation and clines in silviculture

Where the method of cultivation of eucalypts involves planting and allows for a choice in source of seed, the variability which clinal patterns confer is of great importance.

Because this variation is the result of adaptation of the species through the genetic mechanism, it follows that differences found in different parts of a cline will be inherited and that in morphology and physiology at least, the selection of seed from stands in a particular place in a cline may confer a distinct advantage in subsequent silviculture.

Some eucalypt planting is carried out in Australia but very much more is undertaken in other countries. Thus provenance of seed has special importance.

In altitudinal clines, for example, cold resistance will generally increase regularly with altitude although at the upper levels growth rate may fall away. Thus by making enough preliminary experiments in sites to be planted the most suitable seed source—in the above example from which the most suitable growth rate and excellence of form consistent with a designated level of capacity to resist a given low temperature level—may be identified and the silvicultural programme based upon it. Similarly provenance selection may involve the selection of stands as a seed source because they are on particular soil types or display other features which may be inherited which are of silvicultural value.

In spite of distinct advances in silvicultural improvement by various tree breeding methods the choice of the most suitable provenance of a species as a seed source is probably responsible for more silvicultural gain up to the present than any other breeding methods so far employed.

7 Field Distribution and Mycorrhizal Association

7.1 Species population patterns

The geographic range of a species can be represented on a map for most species as a solid block or a few blocks. If the distribution is at all extensive this may cover a substantial range and involve hundreds of square kilometres—but the population is seldom continuous over the more extensive areas. If the map is redrawn with more precision the total population will be seen as a finer or coarser mosaic of smaller segments and if this is further refined by plotting a map based on individual tree location each single unit population may be often not more than a few hectares in extent.

At the same time in several parts of the country the distribution of any single species will be found if drawn on a small scale to be overlapped, partly or totally, by that of one or more other eucalypt species, each of which generally has a distribution partly but not totally coincident with the first species. When the distribution of two or more species which overlap in part of their range is considered in detail, two situations are common—either they overlap completely so that two species occupy the same site; or each forms a mosaic which does not overlap the other at all even though the two mosaics may be generally interlaced, or even very intricately interlaced.

At the same time through wide areas of eucalypt vegetation especially in the drier woodland zones, species occur in pure stands and if there is more than one species in the region each monospecific stand joins at its boundaries another monospecific stand, with mixing of the two species absent or at the most confined to a very narrow junction zone.

For example on the western slopes of N.S.W. the species *E. microcarpa*, *E. populnea* and *E. largiflorens* occur essentially in sequence as one proceeds inland, each occurring on a characteristic site as determined by soil and minor topography.

At the stand boundaries each species occurs on its own site but does not trepass on that of the adjoining species even though a pair of species may occur side by side over many miles along the common boundary between the stands.

If the species concerned are able to interbreed, as is the case with those listed above, interspecific hybrids will be found along and near the line of the stand junction. As explained in Chapter 5 these may be rare, occasional or frequent depending on several factors, but they are seldom entirely absent.

7.2 Mixed stands

In wetter areas where there are generally more species to be found for any unit area—and especially in southern and eastern Australia, the situation is more complex than that described for the western slopes. Very commonly eucalypts occur in stands in which two species are thoroughly mixed (Fig. 7–1) and occasionally there may be mixed stands of three or even four different species.

Where there are mixed stands regularly recurring in areas more or less natural in character and not subject to significant changes due to European settlement, the species which together go to make up the stands are found to belong to different subgeneric groups which do not interbreed. Thus in the south-eastern mainland and Tasmania the sclerophyll forest is made up largely of mixed stands with two species in each, one being of the subgenus *Monocalyptus* and the other *Symphyomyrtus*.

From these conditions a rule may be put forward which is quite widely applicable to eucalypts; this is that in mixed stands composed of two or more species these cannot interbreed, while interbreeding species are confined to separate stands. It must be remembered that this rule is often invalidated by disturbance and its applicability is limited to essentially natural conditions. One can restate it more briefly as: mixed stands are composed of non-interbreeding species; interbreeding species occur in different stands.

The same general pattern is found widely throughout the *Eucalyptus* populations where there are species present from different subgenera (and thus from non-interbreeding groups) (PRYOR, 1959). Mixed stands involving species from *Blakella* and *Symphyomyrtus* are found as in the *E. alba–E. papuana* or *E. alba–E. confertiflora* stands near Port Moresby while in the same area the two species of *Blakella*, *E. papuana* and *E. confertiflora* are allopatric and occupy different sites without ever forming stands with the two species mixed. Examples of mixtures between species of *Corymbia* and *Symphyomyrtus* (*Adnataria*) are common as for example near Grafton, N.S.W. where *E. maculata* and *E. moluccana* form mixed stands; or between *Corymbia* and *Monocalyptus* in the SW. of Western Australia in the stands of *E. calophylla–E. marginata*.

While the breeding behaviour of the sections *Dumaria* and *Bisectaria* in relation to one another is not well known the common pattern in mallee stands in which species of these two sections are mixed in pairs in the same way such as *E. oleosa–E. dumosa* suggests a similar biological system.

7.3 Biological implications

The biological significance of mixed stands in the eucalypts has not been satisfactorily explained. It may simply be regarded as a fortuitous

Fig. 7.1 Dry sclerophyll forest of *E. dives* and *E. mannifera* ssp. *maculosa*.

assemblage of co-adapted species and left at that. On the other hand it is tempting to propose that pairs of species, one each from separate non-interbreeding groups, are in some way biologically complementary and there is a more effective biological unit when both are involved than when they occur alone separately. No satisfactory experimental evidence to help resolve this speculation has yet been presented.

7·4 Mycorrhizae

Whatever hypotheses are eventually developed to explain these distributional features some account must be taken of mycorrhizae in eucalypts and whether or not it is relevant in this regard.

Mycorrhizal associations are general in eucalypts but are more readily evident and have been studied more in the wetter and cooler regions (CHILVERS, 1968a).

The intimate association of fungus and plant roots with the distinct structural units they form was first clearly described in Europe for conifers in the latter part of the nineteenth century. Since then such associations have been described for increasingly large numbers of plants both in Gymnosperms and Angiosperms and for shrubs and herbs as well as trees. In the most intensively studied material, experiments have established the fact that the mycorrhizal equipment may play a part in nutritional relationships in a truly symbiotic way and in pines in particular this is especially associated with phosphorus uptake by the tree and supply of metabolites to the fungus from the tree. At present in eucalypts the role of mycorrhizae is still obscure (UHLIG, 1968) but some of the morphological conditions have been well described.

Ectotrophic mycorrhizae have been found in all eucalypts examined (Fig. 7–2). In this type of mycorrhiza, in contrast with the endotrophic type, the fungus-root develops a characteristic structure. There is an outer growth of densely interwoven fungal hyphae and a penetration of the root cortex as far as the endodermis between the cortical cells by fungal hyphae forming the Hartig net. At the same time there is an elongation of the cortical cells beside the endodermis into a sheath of palisade tissue, which is quite distinctive in transverse microscopic sections.

Several morphological mycorrhizal types have been described, each probably being associated with a different fungus, the precise identity of which is not yet known (CHILVERS, 1968a).

The distinct glossy black type, usually seen in single terminal branches with *Cenococcum graniforme* as the causal fungus, is essentially like the mycorrhizae in other tree genera attributed to this fungus and seems to be similar to that in other continents.

In some other types it is evident from the character of the fungal hyphae that a basidiomycete is involved but in the remainder, the nature of the fungus is still obscure.

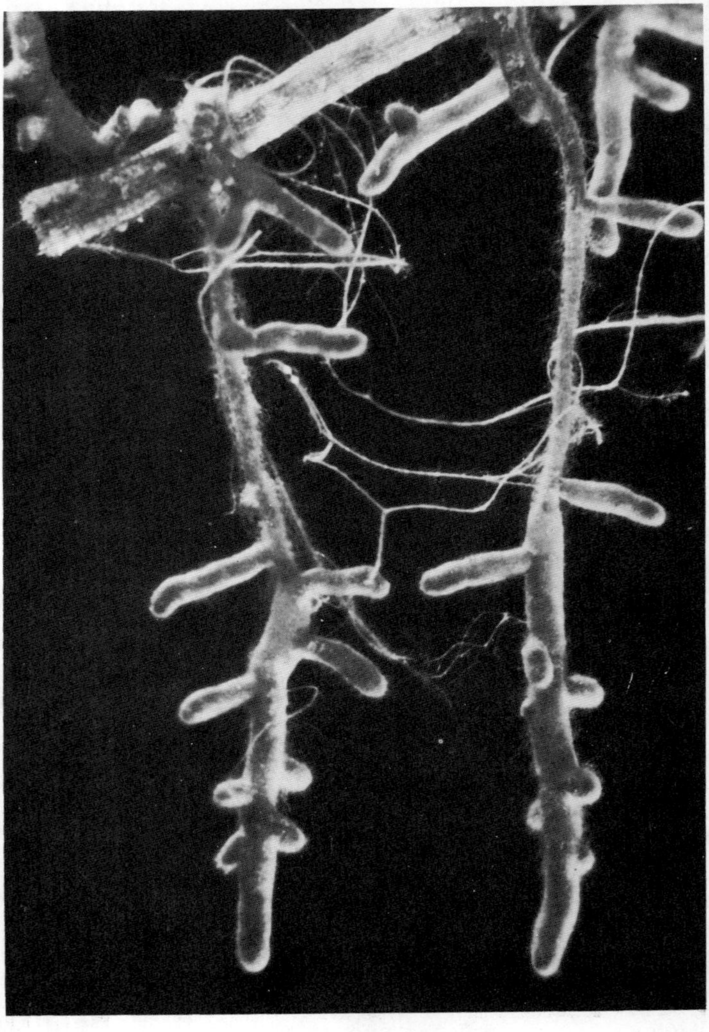

Fig. 7·2 Mycorrhiza of *Eucalyptus*

8 Physiological Aspects

8.1 Drought resistance

Most eucalypts grow in localities where there is marked water shortage for substantial parts of the year. They are plants therefore adapted to drought stress which can be associated either with the dry summers of the predominantly winter rainfall climate of the southern half of Australia or the dry winter and early summer of the northern part of the country where a monsoon type climate prevails and rainfall is confined to the summer months (Fig. 8–1).

Eucalypts are therefore in general xerophytes, belonging physiologically to the group of plants adapted to withstand drought stress. There are several ways in which plants achieve this ability to survive drought, such as the stem succulents as in the *Cactaceae* of the New World or the leaf and stem succulents as in the *Euphorbiaceae* or *Liliaceae* of Africa and Asia.

The eucalypt strategy is different in that the plants develop much hard tissue—sclerenchyma—which confers on them the ability to endure, without lasting damage, severe and permanent wilting. Plants which adopt this form are known as sclerophytes. In a paradoxical way most of the time they do not economize in the use of water but have wide-ranging root systems and an ability to extract water from the soil even though soil moisture tension is higher than that at which more mesophytic plants can extract water. Transpiration rates remain high even though water supply from the soil is dwindling and it is only when severe permanent wilting occurs that there is stomatal closure which inhibits water loss (and of course also prevents gas exchange and photosynthesis) and enables the plant to survive a critical water balance situation for some time.

Not all species are capable of such stomatal closure as GRIEVE has shown by work in Western Australia in which *E. calophylla* stands in contrast with *E. marginata*. The former in a condition of permanent wilting has closed stomata during daylight hours but the latter does not and *E. marginata* continues to transpire even though water lack is extreme and permanent wilting has ensued (GRIEVE, 1956).

No detailed physiological survey has yet been made of the genus to assess which species are capable of the *E. calophylla* type reaction and which are not.

From field observation a few species of northern Australia in the zone of marked monsoonal rainfall with a prolonged drought period often of seven or eight months have become drought evaders and generally become leafless in the last two or three months of the 'dry'. This occurs in

DARWIN
−12°25′S 130°52′E

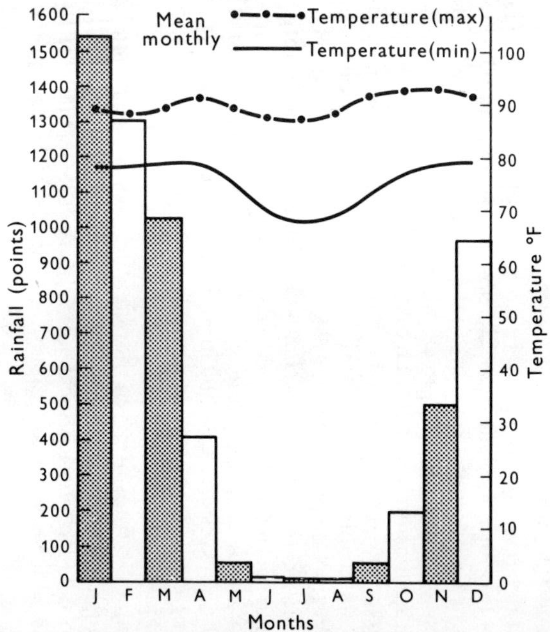

MANJIMUP
−34°14′S 116°09′E

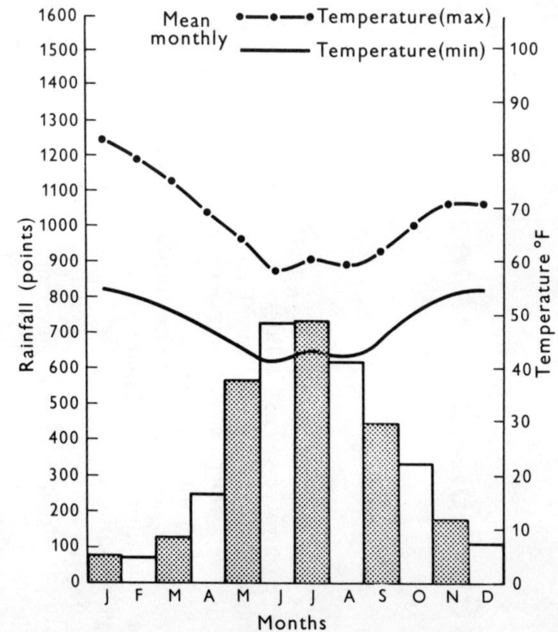

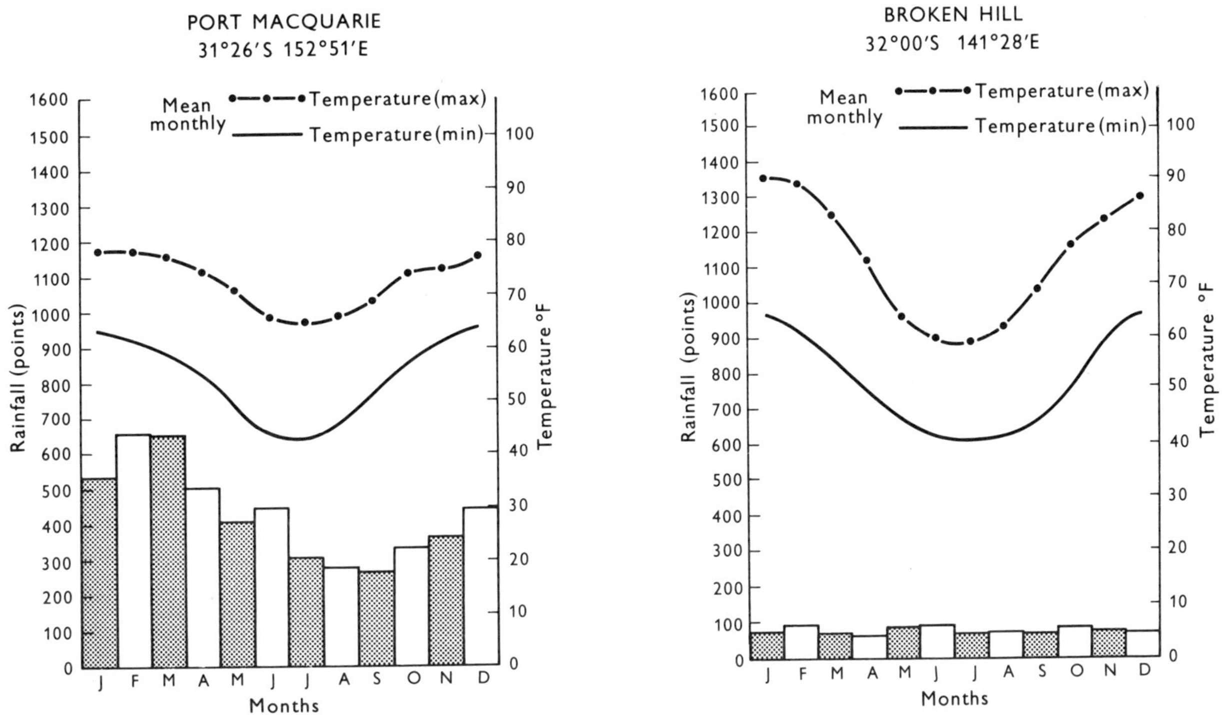

Fig. 8.1 Rainfall and temperature diagrams.

several species of *Blakella* such as *E. clavigera* and *E. grandifolia* as well as in *E. brachyandra*, although this behaviour is often masked by fire which causes defoliation in any case, earlier in the dry season.

8.2 Resistance to low temperature

In the south-eastern mainland and Tasmania the eucalypts are distributed over the mountains and finally form a tree line at about 2000 m on Mt. Kosciusko or at about 1400 m in Tasmania, above which sub-alpine herbfield becomes the vegetation type without any eucalypts being present at all.

They also form a tree line at lower altitudes where topography is suitable to produce temperature inversions and cold air drainage pockets give local 'cold spots'. In this case *below* the tree line in the 'frost hollow' the eucalypts are replaced by herbfield, grassland or fjaeldmark types of vegetation. In both situations the eucalypts at the tree line and also some distance from it endure winter temperatures when the screen minimum is sometimes as low as −15°C and the ground is snow covered for two or three months to a depth of more than 30 cm.

Many woody genera when exposed to such environments have developed the deciduous habit, and even though the species of a genus which occur in lower altitudes or lower latitudes may be evergreen those of the colder zones are deciduous in winter. *Quercus* (Oak) especially as seen in North America in relation to the occurrences in Central America is a good example of this.

The same situation occurs with *Nothofagus* in Australia. There are only three species, two of which are evergreen, but a third, *N. gunnii* which is a species endemic to Tasmania found only in limited stands in the central Tasmanian highlands is deciduous in winter.

This adaptation to cold winters which has been so effectively evolved in the *Fagaceae* is entirely absent from the eucalypts even though *E. coccifera* and *E. subcrenulata* are found side by side with *Nothofagus gunnii* in Tasmania, thriving there in the same climate. In the *Myrtaceae* as a whole the deciduous habit has not occurred at all.

Another requirement of the genus in being able to endure temperatures of −15°C is shown by attempts to grow them in the northern hemisphere in localities where similar temperature readings are made but where there is no regular winter snow cover and, unlike south-eastern Australia, the ground freezes. Frozen ground is lethal to eucalypts and the snow cover regularly present in the natural habitat in the cold parts of the eucalypt environment ensures that they are not exposed to this condition.

From experience with eucalypt species as exotics, another rather special need is disclosed which has been verified experimentally in controlled environment chambers. There are numerous records of eucalypts such as *E. grandis* or *E. tereticornis* being severely damaged by temperatures of only −2°C or −3°C in localities such as the Parana delta of Argentina, even though they are known to suffer little damage in Australia with temperatures which are considerably lower. At more extreme limits such as Texas, U.S.A. or on the eastern Black Sea coast of the U.S.S.R. the most cold resistant species such as *E. gunnii* are severely damaged at times.

A study of the environmental conditions suggests that the capacity to endure without damage any particular temperature—time exposure is closely related to the previous environmental experience of the plants. If there is a progressive and gradual lowering of the minimum diurnal temperature to which the plants are exposed, there is a very rapid 'hardening off' so that after a week of such exposure decreasing the minimum daily temperature by 0.5°C from 0°C, to −3.5°C, the plants develop a resistance to low temperature and will then withstand, −7°C without damage. If they are exposed to a similar low temperature when their foregoing environment has been characterized for the previous growing period of a month or two of no frost and a nightly minimum temperature of not less than, say, +7°C the plants may be killed completely to ground level by a temperature which they will resist if adequately preconditioned.

By using a frost chamber designed especially to study such conditions the rapidity with which physiological adjustment is made in the 'hardening off' process is readily disclosed (PATON, 1972).

The frost chamber is designed to simulate natural radiation frost by using a black ceiling in which temperature is set by liquid nitrogen about 3 m above the test plants to give a radiation temperature gradient of any desired character. The leaf temperatures are monitored by a thermocouple linked to a recorder.

From studies using this equipment the sensitivity of eucalypts to an environment which having been mild suddenly imposes a low temperature period on unconditioned plants is demonstrated and the apparent anomalies which have been noticed even when transplants of species have been made in Australia are explained.

A good Australian example is provided by *E. bicostata* which is quite able to withstand −10°C in Canberra at 600 m elevation in latitude 35°, but when planted at less than 50 m altitude in latitude 29° at Coffs Harbour, New South Wales, it is killed completely merely by a minimum temperature of −3°C which occasionally occurs.

8.3 Effect of photoperiod

Another physiological feature found in many woody species which are spread over a substantial latitudinal range is some photoperiodic regulation of growth and of flowering.

The eucalypt species best suited to such study is *E. tereticornis* which extends from about latitude 8° S. in Papua New Guinea to 38° S. in southern Victoria.

In a species such as *Populus deltoides* (Cottonwood) which extends from Canada to southern Texas in North America, experiment in controlled environment chambers has shown that various growth rhythms and especially the onset of bud dormancy and leaf abscission, are precisely controlled by day length within a given temperature regime and that there has been precise adaptation to this by the species in different parts of its range so that, for example, a provenance from Illinois in latitude 47° N. will become dormant at a day length approaching 14 hours at which photoperiod a provenance from Mississippi (latitude 32° N.) is still able to grow without check until the day length is shortened still further or the temperature drops below the necessary minimum for growth.

In *E. tereticornis* there is no indication of a similar response in different parts of the population, except that a Victorian provenance is able to grow at a lower ambient temperature than a New Guinea provenance, providing the day length is not below 12 hours.

8.4 Nutrition

Some comment is also warranted on the capacity of eucalypts to go through their life cycle on sites of very low nutrient status. Many Australian soils are exceedingly deficient in phosphorus and other essential mineral nutrients. Sites which are like this are characteristically occupied by the peculiarly Australian vegetation of which eucalypts are a part. On the most extremely nutrient deficient sites eucalypts may be excluded almost entirely, but they do occur as small trees or shrubs thinly scattered in most of such localities.

Physiological studies of the mineral nutrient requirements of plants which occur in such sites show them to be in general conservative in their use of phosphorus and presumably other elements.

While eucalypts are clearly able to complete their life cycle and form part of the vegetation of such nutrient deficient sites, they also retain the capacity to respond markedly to improved nutritional status so that when introduced to an environment with higher nutrient levels, especially with better phosphorus and nitrogen status, they respond very markedly by increased growth rates (MOORE and KERAITIS, 1971).

It appears that not only those eucalypt species ordinarily associated naturally with very poor sites respond by increased growth to improved nutrient supply, but that all eucalypts in some degree exhibit distinct response to improved availability of nutrients and especially of phosphorus and nitrogen. This has been shown readily by nutrient experiments under laboratory conditions and in field trials. It is also indicated by the especially vigorous growth response which eucalypts show in their natural environment to the so called 'ash bed' effect (Fig. 8–2). The 'ash bed' response is the growth stimulus which is widely observed in a range of plants in many parts of the world when they are seeded into a site which has been exposed to the effects of burning of a large log. It has been supposed that the stimulus results from the wood ash which is thus generated and added to the soil and while this may at times make a favourable contribution, under Australian conditions experiment (excluding all wood ash) has shown that heat alone on the soil if comparable with that generated by burning logs, produces an identical

Fig. 8.2 The 'ashbed' effect in natural regeneration of *E. grandis*. A. Seedlings on a site under burned logs. B. Seedlings in unburnt soil.

response. Thus the effect (at least in Australia) is really misnamed in being described as an 'ash bed' effect. Microbiological studies of such a site have been made on Australian soil by RENBUSS and CHILVERS (1972) which show that the microbiological recolonization of a soil following such heat sterilization, follows a distinct pattern with bacteria being at first almost exclusively present to which actinomycetes are added after a few weeks while some seven or eight months later after the heat treatment, fungi become a most prominent component of the soil microflora.

The growth response of eucalypts on an 'ash bed' is the same (or even a little more) than that which can be produced on an inert medium by the supply of liberal amounts of phosphorus and nitrogen. But the 'ash bed' response is evidently a complex matter requiring more experiment to elucidate in full detail. It is quite possible that the experiments which produce marked growth increase by the addition of mineral nutrients are simply an alternative way of reaching this level of growth which on an 'ash bed' may be achieved at much lower total levels of, say, phosphorus and nitrogen which may, however, be highly available and at the same time may present a site on which inhibitory micro-organisms (or their growth inhibitory products) have been removed. The 'ash bed' response by eucalypts has been seen in Tunisia, Zambia, India, Maui, Brazil and Texas, to list some examples.

In Australia the greatest contrast between 'ash bed' stimulated eucalypt growth and that on an adjoining non 'ash bed' is seen on the poorest sites, that is on the sites of lowest phosphorus content.

The great sensitivity of eucalypts to soil nutritional status no doubt is related to the nutrient cycling and nutrient balance in natural stands. It is a factor of particular importance too in the cultivation of species of the genus, and in most parts of the world where eucalypts have been planted the desired immediate early growth is realized only if substantial additions of phosphatic and nitrogenous fertilizer are made within a short time of planting.

Eucalypts are evidently able to endure very low levels of many of the minor elements, since they occur naturally and form healthy stands on sites which, where turned over to other crops lead to nutrient deficiency which must be corrected by appropriate supplements. For example, the need to add zinc has been long established when pines are planted on some low fertility former eucalypt sites as in Western Australia or the SE. of South Australia.

Also, especially on soils derived from granite or granodiorite in SE. Australia, clovers used in pasture thrive often only if a supplement of molybdenum is added.

With one notable exception minor element deficiencies are seldom reported when eucalypts are planted as exotics, apart perhaps in the special case of species of *Monocalyptus* which will be referred to in Chapter 11.

The prominent exception otherwise is with the extensive eucalypt plantations in Zambia especially in the 'Copper Belt' where massive applications of boron are necessary in addition to nitrogen and phosphorus initially to secure effective growth.

9 Evolutionary Trends

9.1 General

The concentration of *Eucalyptus* in the Australian zone is *prima facie* evidence that the group has evolved in isolation in the region to produce the great variety of species which are to be seen today. Fossils which are evidently *Eucalyptus* are almost all of very late Tertiary or recent age. The further back in the Tertiary epoch in which fossils are found which are claimed to be *Eucalyptus* the less certain it is that they are referable to the genus. Some fossils collected previously in other continents, for example in Central Europe, have later by study of the cuticular surface of leaves been found to be in quite other families. Unless evidence subsequently emerges to suggest something to the contrary, the situation is that *Eucalyptus*, both living and fossil, is peculiar to the Australian and immediately adjoining regions. Presumably it has developed together with *Angophora* which is quite eucalyptoid in form, largely during the Tertiary period in the environment presented by the Australian land mass through this period.

Compared with the more generalized features of the *Myrtaceae*, *Eucalyptus* has become in both floral and foliar morphology rather specialized, but at the same time certain more generalized features and presumably more primitive features, such as the large and indefinite number of stamens, have remained in the form regarded as basic and characteristic of the family. It is also evident that specialization has occurred unevenly through the species of the genus. For example, the fusion of filaments into four clusters seen in a small number of species of *Eudesmia* is possessed only by that group although *E. microcorys* does show a slight separation of stamens into four clusters of separate stamens without fused filaments which is a development perhaps partly in the same direction. The large majority of species, however, show for the most part a large number of stamens arranged in a band on the staminal ring in the generalized condition for the family, and therefore so far as the *Myrtaceae* is concerned to be regarded as a primitive feature.

9.2 Floral ground plan

In floral structure it is common to find, by analogous comparison with related groups of plants, the development of specializations such as the operculum in *Eucalyptus*. This operculum can certainly be seen as being derived in more than one way, and not always being derived from the same structures. In the most common case where there are two opercula, one evidently being made up of fused sepals and the other of fused petals, these are readily interpreted as a specialization which is a feature of the group and one which has been derived in the evolutionary sense by fusion of parts from a simpler forerunner in which the sepals and petals were separate and free, as they are in many other members of the *Myrtaceae*.

Likewise the inflorescence can be seen as a cluster of flowers brought together by condensation of a branching system which in the more general condition is a branched cyme as is seen in other living genera, for example in the closely related genus, *Angophora*. The condensed arrangement which produces a cluster actually, but not at once evidently, cymose in structure and which is quite reasonably described as an umbel, is again an example of specialized floral development in the genus.

The same is true of features of the seed as in the seed coat and raphe; this latter being the vascular strand running from the hilum to the chalaza which is in a great many species modified from the more primitive and generalized type. The general type is considered to be the anatropous ovule where hilum and micropyle lie closely side by side which is a feature of *Monocalyptus*, and not the hemitropous ovule which is found in most other members of the genus. This suggests that specialization in some groups has proceeded in some organs, whereas it has not occurred in some others and the more generalized primitive characters have persisted in those.

In foliar organs the same thing can be found. The more primitive and general type of leaf is, by comparison with other genera in the *Myrtaceae*, considered to be a dorsiventral leaf with an opposite decussate phyllotaxy. This is the characteristic morphology of many of the juvenile leaves in the eucalypts and especially so in the strikingly heteroblastic species, whereas the common form of adult leaf has a modified phyllotaxy which nevertheless can be seen as being clearly derived from the general parental form. The adult leaf has additionally a condition in which the leaves have more or less equal sides with stomata on both surfaces so that they are isobilateral and not dorsiventral as is the primitive condition and as seen in the majority of *Myrtaceae*.

Again, not all eucalypt species show this kind of adult leaf development and particularly in the section *Transversaria*, a cluster of well known species which has leaves which, though sclerenchymatous, nevertheless still retain the hypostomatous arrangement, unlike that of the majority of species in the adult phase. There are some too which retain the juvenile leaf morphology right through their life, which again may be taken as a preservation of a feature more primitive than most species display.

Evidence of development in other floral features can be seen in the specialized adnate anthers of the section *Adnataria*. The two-celled versatile anther with dehiscence by long parallel slits is the general myrtaceous condition. But this is not preserved in any of the *Adnataria* where the versatile condition has been lost in favour of an adnate anther and the dehiscence by small pores or very short slits. As a further example a few species of *Monocalyptus* in Western Australia did not follow the change by confluence of the anther cells characteristic of the majority of species of *Monocalyptus*, which is another variation and presumably specialization, of the anther. But species such as *megacarpa* and *preissiana* retain the generalized morphology in this regard, although in other respects they are clearly in the subgenus *Monocalyptus*.

9.3 Rate of evolutionary change

An estimate of the way in which evolutionary changes have occurred with time is given by making a comparison of the eucalypt species of the extreme south-west of Western Australia and those of south-eastern Australia. This, in many ways, resembles comparisons which have been made between genera having pairs of species, one in North America and one in Eurasia. There are many general habitat features in common between these two Australian areas and they are both rich in species characteristic of the Australian element in the native vegetation. It seems that the two areas must have been isolated from rather early in the Tertiary period and at least since some time before the Miocene. This is inferred because the large Miocene limestone deposits characteristic of the Nullarbor Plain adjoining the Great Australian Bight between Western Australia and South Australia seem likely to have been a barrier to the interchange of species from the south-east corner and the west corner of the continent, even when the connecting area became dry land after the Miocene period. In the two regions similar major systematic groups are present, for example in Western Australia there are a couple of species in the *Transversaria* one of which, *E. diversicolor*, is evidently reasonably closely related to species in the same section in eastern Australia, such as *E. saligna*, *E. grandis* and *E. propinqua*, yet it is distinct from all of these and is also distinct in its physiological behaviour. In the physiological aspects the eastern species are suited to a uniform or summer rainfall regime whereas the Western Australian species is from the evidence of trees planted as exotics, very clearly only at home physiologically in areas of winter rainfall which is characteristic of the Western Australian climate. It has, for example, grown well in the area near Ovalle in Chile and at Tokai near Capetown in South Africa. There is a similar comparison between *E. calophylla* and the pair of eastern species *E. gummifera* and *E. intermedia*. All three species are in the subgenus *Corymbia* and are obviously related one to the other, although again *E. calophylla* is perhaps less like other species of *Corymbia* in eastern Australia than some of those are like each other. There is a further similarity as between western *E. marginata* and *E. patens* and eastern Australian species of *Monocalyptus*, but here the separation in morphology is perhaps more extreme than in the former two which could imply that *Monocalyptus* has been separated for a longer period than species in other subgenera. Nevertheless, in Western Australia three of the major subgeneric groups, *Corymbia*, *Monocalyptus* and *Symphomyrtus* (*Transversaria*) are represented by a few species which produce a situation parallel to that in south-eastern Australia. This similarity is extended to the pattern of species distribution in mixed stands and both in interbreeding capacity and barriers to interbreeding which have been described for the species of eastern Australia. Especially because of low topographic relief, habitat diversity in Western Australia is much less than eastern Australia, and it seems that the opportunity for speciation as it occurred in the east was not present to the same extent in the west and the situation is simpler in that fewer species are involved. But the characteristic pattern remains the same and the morphological divergence of the species in each group from those of eastern Australia in comparable groups is consistent with there having been a substantial period of isolation which has lead to a degree of independent evolutionary divergence.

There is a further point of interest between east and west. The sclerophyll forest communities in both parts of the continent are on soils of low nutrient status and often on old land surfaces, frequently remnants of a lateritic profile. On the inland slopes of the ranges in New South Wales and Victoria and on the inland side of the Darling Range in Western Australia there are extensive woodlands of eucalypts on soils which are often younger and in general a little more fertile than those supporting sclerophyll forest. A separate suite of species is present, one east and one west, occupying the range of sites, often a consequence of soil differences in these general areas. In eastern Australia the species involved are predominantly from the section *Adnataria*, that is Boxes and Ironbarks which are a distinctive feature of this region. In Western Australia the species involved are to a large extent from the section *Bisectaria* and to some extent *Dumaria*. There is very little interchange between east and west in species of these groups. There are only a handful of species of *Bisectaria* in very isolated occurrences in eastern Australia such as *E. pachycalyx*, *E. squamosa* and *E. bakeri*, and in the reverse way in the more arid parts of Western Australia one or two species of *Adnataria*, but in the main the two groups are distinctly separated in their respective geographic zones.

9.4 Evolutionary convergence

The woodland habitat is quite similar in both areas and there are some striking examples of convergent evolution in which pairs of species of different systematic groups, one found mainly in the east and one in the west, have each developed a form which makes them in appearance strikingly similar. For example, *E. cambageana* (*Adnataria*) of Queensland resembles very considerably *E. kondininensis* (*Dumaria*) of Western

Australia; *E. gomphocephala* (*Bisectaria*) of Western Australia has many of the features that are found in *E. microcarpa* (*Adnataria*) of eastern Australia and *E. salmonophloia* (*Bisectaria*) Western Australia, has general aspects which resemble *E. thozetiana* (*Adnataria*) of Queensland. There are numerous examples in *Eucalyptus* of convergence which are not associated with the separation of the two groups referred to in eastern or Western Australia. In single characters this is common. For example, the decorticating bark giving a smooth trunk is characteristic of at least one species in all major systematic groups, for example, *E. maculata* in *Corymbia* is smooth barked, many species of *Symphyomyrtus* are the same, such as *E. viminalis*, and in *Monocalyptus* there are *E. tasmanica* or *E. haemastoma* and again *E. leucoxylon* in *Adnataria*. If one looks at the absence of lignotubers, this is found to occur in most groups, there being several species in *Monocalyptus* such as *E. regnans*, *E. delegatensis*, *E. fastigata*; several species in *Symphyomyrtus* such as *E. diversicolor*, *E. grandis*, *E. camaldulensis*, *E. dunnii* and *E. nitens*; some in *Bisectaria* as *E. astringens* and *E. gomphocephala* but none so far is known in *Corymbia*, *Blakella* or *Eudesmia*.

The presence of the three flowered inflorescence has a similar pattern, being rare in some subgeneric groups and common in others, for example in *Monocalyptus* it is found probably only in *E. triflora* and in some populations of *E. coccifera*. It is common in *Symphyomyrtus* as in *E. rubida*, *E. bicostata* and *E. caesia*, occasional in *Adnataria* as in *E. leucoxylon* and in *Corymbia* as in *E. maculata* and in *Eudesmia* in *E. odontocarpa*.

It is evident that similar trends of specialization have been followed independently at times in groups of species which are otherwise quite different and certainly derived from different evolutionary lines; in short there has been evolutionary convergence.

9.5 A monophyletic or polyphyletic genus?

It remains an open question as to whether all species now known as *Eucalyptus* have been derived from a single proto-eucalypt and are then monophyletic, or whether they have emerged from more than one prototype in the *Myrtaceae* and are thus polyphyletic. The morphological and genetic differences between the major subgeneric groups are substantial and perhaps reflect a diverse origin but there is at present inadequate information to conclude these speculations.

The view may be taken that if the group is polyphyletic then what we now call *Eucalyptus* should be divided into more than one genus. If on the other hand it is monophyletic then it stands as one genus with perhaps the addition of *Angophora*.

For many purposes, there is little loss of information if one, in recognizing several subgenera, chooses to retain them at this ranking or to elevate them to separate genera. Perhaps the most telling reason for retaining the genus *Eucalyptus* is the purely pragmatic one of avoiding a highly complex series of name changes which would be necessary in terms of the International Rules of Botanical Nomenclature to conform to them.

Since to reach such a decision there would always be a highly subjective element involved, and since little gain could be made in conveying information, many botanists (but not all) would prefer to let the matter rest until, if ever, new elements can be brought into the consideration.

10 Fire and Settlement in the Eucalypt Environment

10.1 The fire environment

Eucalypts as part of the general sclerophyll vegetation each year endure periods of water stress which usually are associated with inflammability. At the wet climatic limits where the trees abut or are undergrown with rainforest, high inflammability may be at only relatively long intervals of some years, while at dry limits production of fuel from dry plant parts may be so sparse that only after exceptional rainfall is there enough fuel from quick growing grasses or other herbaceous plants to carry a running fire and therefore burning again is quite infrequent—often only at intervals of some decades. Between these extremes lies the great bulk of the eucalypt vegetation, much of it with a grassy undergrowth (Fig. 10–1) which is capable of burning annually and the remainder becoming dry enough mostly each year to burn when sufficient fuel has accumulated

Fig. 10.1 Savannah woodland in North Queensland near Mt. Garnett. The tree is *E. drepanophylla*, an Ironbark.

from dry plant parts such as leaves, twigs and bark. In this latter case burning is generally possible at two- to three-year intervals.

Fire, in this case, means a fire which is spontaneously propagating once having been initiated from a primary ignition point. The rate of spread and intensity of burning is related directly to temperature, wind, fuel load and humidity. Throughout its history, *Eucalyptus* must have occupied an environment in which fire has been present. Before the arrival of man—known at present from radio-carbon dating to extend back at least 30 000 years—fires were ignited by lightning. This still regularly occurs at various frequencies in different parts of the country. Occasionally other natural causes such as volcanic action can be imagined as having also lead to ignition but this would be very rare.

At the time of arrival of aboriginal man the eucalypt vegetation was already certainly a fire adapted type in which the plants, both eucalypts and others, had already evolved features which made them well suited to an environment in which fire was a regular feature, although this would be almost all dependent on lightning for ignition.

Aboriginal man must have made a significant change and it is likely that he developed some distinct patterns of burning, although the nature of these is scarcely more than hinted at by evidence from different sources such as the vegetation itself and the rather scanty historical records. It is likely that in most of the eucalypt vegetation, but especially in the more southerly parts of the country, fire incidence with aboriginal man present became much more frequent and probably generally less intense for the most part. Almost certainly there was in any one region a mosaic of areas burnt in different parts of the season in different years. In Northern Australia it is likely too that the arrival of the aborigine meant more frequent but perhaps less severe fire.

After European settlement to judge by vegetational evidence, especially in the more southerly parts, there was yet another change, for fire probably became rather less frequent but often more intense than in the aboriginal period. In northerly regions it was probably less frequent too, but perhaps there was not much change in intensity. A clearer picture may yet emerge.

10.2 Eucalypt resistance to fire

The eucalypt response to fire is readily seen and it is dependent both on intensity of burning as well as the intrinsic characters of the trees themselves, since not all respond in exactly the same way.

The great majority of eucalypts when well grown trees are very resistant to fire. At times a fire can be so intense, due to high fuel loads and extremely suitable meteorological conditions, that all the leaves and small branches and even branches up to say 10 cm diameter may be completely burned off. The outer bark may be completely carbonized but at the same time the bark is so thick and such an effective thermal insulator that the cambium and concealed buds are not damaged and are able to sprout, forming epicormic shoots over the trunk and surviving branches. In species which are strongly heteroblastic these epicormic shoots have juvenile leaf morphology but after 2–3 years this is superseded by normal adult foliage and normal flower development and seed production follows. A little later it is often not easy except by careful scrutiny to recognize that a given specimen has had this history of fire and has responded to it by what amounts to complete recovery as a functional organism.

If the fire has been less intense leaves only may be scorched, partly or completely, die and fall to be replaced as soon as growth recommences, by a crown of new adult leaves.

If the burning is still less severe a large tree may be essentially unaffected by it, but only small seedlings may be killed completely. Young coppice shoots from a lignotuber may be killed completely but as soon as seasonal conditions are favourable a new shoot develops rapidly from the same lignotuber. It is evident that lignotubers greatly facilitate survival and are very effective in aiding stand recovery after fire.

Vegetative recovery, either by epicormic shoots or regrowth from lignotubers, is the general survival method of most eucalypts after fire. But in addition a hot fire will induce from ripe capsules a general shedding of seed which otherwise would fall sporadically as the branchlets carrying the fruit die. Thus another response to heavy fire is a heavy seed rain, which being associated with favourable site conditions as is generally the case when litter and tops of plants in the shrubby undergrowth are burnt, is accompanied by a good deal of regeneration of new seedlings too.

While most eucalypts behave in this way a small number of species has an alternative strategy. One group of particular interest is made up principally of some species of *Monocalyptus* and a few of *Transversaria*. These are those such as *E. pilularis* and *E. grandis* of N.S.W. and south Queensland, *E. sieberi* of coastal sites of south N.S.W., Victoria and Tasmania, *E. regnans*, *E. fastigata*, *E. delegatensis* and a few related species mainly of the south-east mainland and Tasmania and *E. diversicolor* of south-west Western Australia.

These species in general are rather less resistant to fire in the adult stage than the great majority of eucalypts. One of the most susceptible, *E. regnans*, is often completely killed by an intense fire and even though a large tree, may develop no epicormic shoots. This and similar species also have no lignotubers so that young seedlings or treelings are especially susceptible to and sustain lethal damage even from light fires.

There is a compensating condition. These species in most years carry heavy crops of fruit from which the seed rain is very intense after a heavy fire. Because also they are essentially species of wet sclerophyll forest the removal of undergrowth species and killing of trees of various ages according to fire intensity results in a particularly favourable situation for immediate seedling regeneration. It is general in those species for a 'wheat field' regeneration of 20 000 or more small seedlings per ha. often to follow intense fire. The wet sclerophyll forest is a vegetation type in which fires are usually intense if they do occur because in the majority of years conditions are too moist for a running fire and it will ignite and produce such a fire only if there are high winds, low humidity and high temperatures. These conditions always lead to an intense fire and this in turn produces very favourable conditions for dense regeneration by seedlings.

It is clear therefore that this group of species is adapted rather differently from the large majority. A consequence of this behaviour is that extensive regeneration of, for example, *E. regnans* has occurred in the past before European settlement, evidently according to times of severe fire incidence. Stands of substantial size tend to be even aged and of a different age from adjoining stands by rather large time intervals of 50 years or so, especially in the period from about 200 to 500 before present.

10.3 Eucalypt susceptibility to fire

An example of excessive fire sensitivity of a species in an environment where fire does not exist at all is given by *E. deglupta* which occurs in a pioneer phase leading by succession to rain forest in the wet tropical lowlands such as in New Britain and Mindanao. The prevailing vegetation is rain forest in which fire is virtually unknown.

E. deglupta which has the thinnest bark of all eucalypts—a mere 3 mm—is found to be exceedingly fire susceptible and will die from the effects of the tiny fires used in clearing for gardening under shifting cultivation or even a light grass fire which may occur in areas where the tree has been planted away from its usual habitat. In this respect it is quite as fire sensitive as species of *Populus* or some of the more sensitive conifers of the northern hemisphere.

There is evidence that the regeneration of Mallee is also related to fire. It is quite remarkable to note the extent to which seedlings are absent from older stands and to observe the dense regeneration which follows a fire. However, fire in Mallee is rare. There is in most years not enough fuel on the ground to carry a self-propagating fire. Occasionally there is enough seasonally extended rainfall in the Mallee zone where the total average annual precipitation is from 200 mm to 300 mm a year, to result in substantial grass and herbaceous growth which when dry later in the year is inflammable and fire can then occur. The Mallee species are generally quite fire sensitive and are usually extensively killed after this happens. The following seedling regeneration is prolific as well as being accompanied by some regrowth shoots from the large underground lignotuber characteristic of Mallee species.

In the woodlands with a grassy undergrowth, which are very widespread in Northern Australia, fire is a relatively minor event. Because of the monsoon climate there is annually enough fuel to carry fire but it cannot accumulate sufficiently to produce a heavy fuel load, even without an annual burn. The result is that any fire is relatively light and while it may be heavy enough, according to the point in the dry season at which it occurs, to cause complete defoliation, this damages only leaves and small twigs on bigger trees and often only smaller treelings or even seedlings. Annual fire is a common pattern in the tropical eucalypt woodland.

In the extensive areas of dry sclerophyll forest, which are found so widely in eastern and southern parts of the land, fire is a periodic event and apart from the wet sclerophyll forest there is little that is not burnt once in 20 years and much is burnt every 3–5 years. This has certainly been the pattern since aboriginal occupation.

10.4 European settlement effects

European settlement of Australia which began near the end of the eighteenth century involved, so far as land use was concerned, the introduction of crop growing and grazing of domesticated animals, both of which practices were unknown to the aborigines.

Progress in developing these farming practices was at first quite slow and land clearing—which generally meant the removal of eucalypts—did not cover a very big area even at the end of the first 50 years. But by the middle of the nineteenth century and up to the present day there have been increasingly large areas devoted to these uses. Not only eucalypt vegetation has been affected but also some land originally occupied by other vegetation types has been much changed—as, for example, the substitution by pasture or crop plants of quite extensive areas of rainforest in New South Wales and Queensland and shrub steppe of *Atriplex* and *Kochia* in New South Wales, Victoria and South Australia and extensive heathlands in Western Australia.

The most evident effect on eucalypts has been where clearing of trees has been undertaken to permit cropping—particularly for wheat and other grain—and when there has been similar tree removal for the establishment of cultivated pasture for sheep and cattle. Very large areas of eucalypt woodland in the southern half of the country have been transformed in this way and the total eucalypt population in the vegetation types concerned has been much reduced. Large areas of Mallee have been completely cleared for wheat farming too, and some sclerophyll forest has been converted to pasture especially by the addition of mineral nutrients to the soil after tree clearing to favour the growth of pasture grasses and clovers.

There has also been at a lower level of change a less evident but equally widespread effect resulting from extensive grazing mainly by sheep and cattle, of large tracts of land. Much of this effect is associated with the use of fire to make sclerophyll forest more productive for grazing by cattle and to some extent, sheep. The main effect on eucalypts in this case is determined by the fire regime which in turn affects tree regeneration patterns and stand characteristics.

Where woodlands have been intensively sheep grazed without substantial tree removal the change from the original marsupial grazing population to sheep grazing has been associated especially with a substantial elimination of seedling regeneration and so a potentially non-viable condition for the eucalypts—unless the grazing management is modified to permit more seedling regeneration survival to replace moribund trees as they die. Eucalypts though mostly long lived plants—200–300 years is usual—are not immortal and eventually become moribund and die.

Complete removal of eucalypts has followed many other activities, such as the less extensive cropping of many other agricultural crops besides grain like fruit, vegetables, tobacco, grapes and silvicultural crops like pines. But of significance too is the tree removal due to the spread of cities and the development of communications, especially roads and power lines as well as removal due to various types of mining.

10.5 Eucalypts and the conservation of species

The change in the plant cover of Australia during the last 100 years has been very great and a question to consider so far as eucalypts (and other plants) are concerned is whether any species has been eliminated or is in danger of extinction. If so, a further question arises as to whether effort to preserve them is warranted.

The merit in conserving biological species has been widely canvassed. While most botanists would no doubt favour the preservation of all plant species, at least by some examples still in a natural site, the argument that opposes the view that conservation of all species in a natural habitat at *any cost whatever* must *per se* be justified, cannot be ignored.

Biologically, the history of life is one of continual replacement of one species by another and the sequential extinction of species in time. The big difference in this century compared with the past is the rate at which such changes are occurring and the extremely heightened effect of man in generating this rate of change.

The preservation of species has merit to the scientist because it retains living material in an array for study and experiment. This array is the product of evolution and once lost can never be recovered.

For the layman the retention of species will often mean ensuring the availability of organisms to be seen and studied and so provide whatever aesthetic or intellectual satisfaction this means to anyone with the urge to explore living material.

For those who see man's future as highly dependent on the success with which man can manage his environment, then the preservation of species, and in our case here of eucalypt species, means the preservation of an irreplaceable gene pool to serve man in the future.

The present situation with eucalypts is that there is no record of any species or subspecific group having been lost. In as much as each individual or interbreeding population may be considered unique, just as every individual man may be considered unique because he is genetically so, there have been some losses because there have been changes which have eliminated whole populations, but at the present moment very closely similar populations are extant. Those species which are most reduced and likely to be reduced further are those of limited extent and occupying habitats where current land use tends to replace them. For example, *E. camfieldii*, always a species of restricted range and tiny total population, mostly on sandstone in the northern suburbs of Sydney and a little to its south, has been reduced considerably but still survives on sites too rugged for urban development. Here it is the urban spread which generates the risk. Likewise, *E. curtisii* occurs mostly on sites in the path of urban development near Brisbane. But in many cases the survival of restricted species is ensured because the natural features of the site have already led to their being preserved. For example, *E. mitchelliana* which is known only from the northern end of Mt. Buffalo in Victoria is so protected and *E. alpina* is in the Grampians National Park in western Victoria.

Widespread species, even those occupying sites now mainly devoted to agriculture, are still to be found in large populations along roadsides, in reserves, state forests and the 'back paddocks' of farms and the like, but the situation can as yet scarcely be called static.

A main concern is to ensure that the still rapid rate of change does not engulf all the existing populations of any species before adequate reservations have been made. This is urgent with highly restricted species in areas where change is likely and also for those species characteristic of areas occupied by vegetation types already much devoted to land use which largely eliminates eucalypts. It is relatively well cared for in eucalypt sclerophyll forest under existing land use patterns and for those species which make up the main eucalypt wood producing forests.

11 Eucalypts as Exotics

11.1 Historical

Slight botanical contact had been made in Western Australia from the time of Dampier but it was not until European contact in eastern Australia was made first in 1770 by Banks and Solander, who accompanied Captain James Cook, that some of the many distinctive native Australian plants became known and interest in them was awakened. In the following decades various species, amongst them eucalypts, were introduced and grown in botanic gardens throughout the world so that by the middle of the nineteenth century some eucalypts were well known ornamental trees in those European and other gardens where the climate was mild enough for them to survive. In this way Tasmanian Blue Gum (*E. globulus*) became widely known and planted. For example, it had already been established in Mauritius and was then taken to Capetown by a Governor who was transferred in 1836. The oldest tree near Ootacamund in the Nilgiri Hills in India was planted in 1863 and the species had been introduced and flourished in Portugal and Spain by the same time. By about 1900, growing world population and increasing timber demand turned attention to the possibility of using eucalypts for timber production, especially for sawlogs. There are deficiencies in the genus for this latter purpose, especially when attempts are made to grow them very quickly and to harvest them at an early age of say 15 years or even less and in processing them by traditional methods. This poor performance as a sawlog when young resulted in a slump in enthusiasm for planting species of the genus for timber production, although various species remained in cultivation and became widely used for shelter and ornament.

By 1945 there was a further change due to the growing demand for wood for pulp and paper as well as other reconstituted uses such as hardboard which, together with technological changes which made the use of eucalypt timber possible as short-fibred pulp, have led to the genus becoming one of the most widely planted at the present time in the warm temperate and tropical parts of the world.

11.2 Insects in relation to growth

The outstanding single feature of the genus is its capacity for rapid growth as an exotic, if soil and climatic conditions are generally suitable. One of the reasons for this fast growth is that in their adopted habitat there is usually a lack of the insects which are so prevalent on them in their natural Australian conditions. There is specific to the genus, a very large number of insects which live on eucalypts, many of them being leaf-eaters such as the scarabs, chrysomelids, curculionids and tenthredinids. There are also sucking insects such as the psyllids, jassids and coccids as well as wood borers like the cerambycids and the large cossid moths. Almost always eucalypts have been introduced to countries outside Australia by seed and this means, fortunately, that the insects are left behind. There are two important consequences of this, firstly, because the leaf-eating insects are absent, the crowns remain bigger, with more leaf area and growth enhanced accordingly. Secondly, if by chance an insect which feeds on *Eucalyptus* is introduced as occurred with the curculionid, *Gonipterus scutellatus* in South Africa—it is likely to appear without its natural predators and so become epidemic in its defoliating effect. In the South African case this has been controlled essentially by the subsequent introduction of a parasitic wasp, also indigenous to Australia which has restored a reasonable balance and the damaging insect is no longer of much economic importance. There are a few similar cases in other parts of the world such as the longicorn, *Phoracantha* in the southern and eastern Mediterranean and the chrysomelid, *Paropsis* in New Zealand. There is little doubt that the fast growth of *Eucalyptus* as an exotic is partly because of its freedom from the Australian indigenous pests which are so much a feature of it in the natural habitat; but there is another aspect too. Eucalypts occur in sites in Australia which are for the most part deficient in plant nutrients and especially so in phosphorus; a characteristic of the region as a whole. In many areas where they are planted outside Australia the basic fertility levels are higher than in the natural Australian habitat. Although the eucalypts are capable of growing and going through their life cycle under conditions of low nutrient status they nevertheless are very sensitive to added nutrient and respond very markedly in growth if nutrient supplies, particularly of phosphorus and nitrogen, are increased. It is for this reason also that in many cases growth is greater when they are seen as an exotic rather than under natural conditions. In any case too, the practice adopted in growing eucalypts which very often involves soil cultivation and the addition of fertilizer causes more rapid early growth than is often the case with natural regeneration. Some growth increase with these techniques results in Australia too—but unless insects are controlled the growth rate falls short of that common in similar climatic conditions overseas.

11.3 Limits of introduction

11.3.1 Climatic

Within Australia the environment is one in which *Eucalyptus* for the most part is continually present throughout the range of the genus and each species occupies a relatively circumscribed, more minor habitat, adjoining other habitats which are likewise occupied by other species of eucalypt. If one imagines a situation in which only one species of *Eucalyptus* were present it seems feasible, because of experimental transplant evidence, that it would be able to occupy a much larger suite of sites than it does when other species are present. That is to say, there is evidence that the ecological amplitude, in terms of habitat which a given species is capable of occupying if the competitive effect of other eucalypt species were removed, is generally much greater than that which its site occupancy suggests in its area of natural occurrence. This means that in making introductions the natural range of the species is only a partial guide to what may be achieved as successful growth in the various conditions in which it might be planted. This is particularly the case with species which are naturally restricted to rather special sites, such as, for example, the water courses and flood plains occupied by *E. camaldulensis* or the somewhat swampy or even at times partly saline estuarine sites in which *E. robusta* is frequently seen. It appears in the latter case for example, that *E. robusta* occupies such sites because it is able to grow under these rather adverse conditions whereas species growing on adjoining but better drained areas cannot endure them. At the same time the species on the better sites are able to exclude *E. robusta* if placed in a competitive situation with it on those sites. Thus, the two species, *E. robusta* and *E. camaldulensis*, though confined to sites of limited ecological amplitude are found to be very adaptable and are planted successfully as exotics over a very wide range of site conditions, which generally are considerably wider than those in which they occur naturally.

The main limitation in planting *Eucalyptus* outside Australia is by low temperature. The species of the colder areas of Australia (particularly Tasmania) grow in sites which receive regular snowfall so that the ground blanketed each winter by snow, never freezes and the soil temperature remains just slightly above 0°C. Whenever attempts are made to plant eucalypts in sites where the ground freezes in winter, such as continental Europe, Asia and North America they are inevitably followed by death and there are no survivors. Thus, in general *Eucalyptus* is limited to about latitude 45° North in Eurasia, the exceptions being especially those parts of the United Kingdom and Ireland and the north-west coast of Portugal where because of the oceanic influence conditions remain mild. In this special case eucalypts have grown quite successfully even as far north as Argyllshire in latitude 56°. The species which thrive in these circumstances are some of the Tasmanian endemic species such as *E. urnigera*, *E. johnsonii* and *E. gunnii*. To some extent a similar effect is to be seen on the west coast of North America but it is much less marked because of differences in the ocean temperature on the Pacific side of North America compared with the United Kingdom side of Europe. At the other end of the scale the main limitation to *Eucalyptus* growing is in the low altitude, wet tropics. The only species which occurs naturally under these conditions, in latitudes of less than 10° and with rainfalls of 3000 mm or more, is *E. deglupta* which thrives in these circumstances. But other species planted there do not display satisfactory growth unless they have the advantage of at least 600 m of altitude in such latitudes. Between these two limits of the wet, lowland tropics and the colder conditions associated with higher latitudes or very high altitudes, *Eucalyptus* will grow on a wide range of sites and has been cultivated on many. They are particularly well adapted to areas where the climate is periodically dry, such as in the monsoon parts of Asia and Africa.

11.3.2 Soils

With regard to soil type *Eucalyptus* is rather deficient in providing species capable of enduring highly calcareous conditions or heavy textured soils with very high pH values, and perhaps with gypsum present. Thus, those few species which do thrive on limey soils or heavy textured ones have been widely planted and are much valued. One of the outstanding is *E. gomphocephala* (Tuart) which occurs naturally on calcareous sands and other calcareous sites on the coastal strip of southern Western Australia, mainly south of Perth, and it has proved very effective in plantations around the Mediterranean both on the European side but especially in North Africa in countries such as Morocco, Algeria, Tunisia, Libya as well as Israel, Syria and Jordan. A species which resists both these conditions and also very heavy clays and a low rainfall is *E. microtheca* (Coolibah) which has been very successfully grown for wood lots in the cotton growing Gezira zone of Sudan. The annual rainfall is so low (400 mm) and of such short duration here ($2\frac{1}{2}$ months) that some irrigation is necessary for reasonable growth. Water supply can be phased in such a way that enough can be applied when the peak water demand for cotton farming has passed.

11.3.3 Rainfall regime

There is a rapidly growing body of information about the performance of eucalypts planted as exotics, but it is still rather early to come to the general conclusions which may later be possible, in providing rules which summarize introduction possibilities. Nevertheless, some pointers are available. It is noted already that species which are characteristic of the winter rainfall zones of Australia, particularly south-west Western Australia and South Australia, grow poorly if planted in a climate where there is predominantly summer rainfall, but the reverse is not the case. Those species which grow naturally in areas of summer rainfall often will perform in cultivation quite well in areas of winter rainfall, thus *E. citriodora* is a successfull ornamental tree in Perth and Adelaide, even though it is confined to Queensland naturally and does not occur south of about latitude 26° S. there.

11.3.4 Systematic group

There is a general observation that species of the subgenus *Monocalyptus* seldom perform well outside Australia, and are especially noted as failures in those warm temperate or tropical areas where there is a distinct dry period. The main exception to this observation is the performance of a good many species of this group in the Republic of South Africa and to some extent in Rhodesia, Kenya and Tanzania. No precise experiment which explains the commonly experienced failure of this group of species has yet been presented. It appears to be a nutritional disorder and is the sort of thing that is associated with lack of a suitable mycorrhizal fungus, at least in other genera known to depend upon the mycorrhizal situation. There is no experimental verification that this is the cause of growth deficiency and the position is made more complex by the occasional success seen in limited areas from time to time with some *Monocalyptus* species. The possibility that mycorrhizal conditions are involved is enhanced by the general observation in southern Africa that eucalypt plantations, whether of *Monocalyptus* species or not, are found almost always to have gasteromycete fungi present and in particular fruiting bodies of *Pisolithus* can usually be seen under the trees.

11.3.5 Genetic degrade

One of the very common consequences of the introduction of *Eucalyptus* species, often in trial collections or arboreta removed far from their natural habitat to areas where they are exotic, is the opportunity which is thus presented for interspecific hybridizing between species which in the natural circumstance in Australia have not been exposed to one another in such a way that hybridizing could occur and the products from it grow into mature plants. It is both a matter of eliminating geographical isolation and of modified flowering time behaviour, if not of other factors.

One of the outstanding instances is in Brazil where the species *E. 'urophylla'* from Timor was introduced by seed in 1919 at the famous arboretum created by Navarro de Andrade at Rio Claro in São Paulo. There it develops a characteristic which is rather unusual for eucalypts in their natural environment, having a very long flowering period. Under the conditions in which it is growing in Rio Claro it is at anthesis, to some extent, for ten months of the year. Beside it, planted at about the same time, is a plot of *E. tereticornis* which is one of the few eucalypt species which is, at least in part if not completely, wind pollinated. There have thus been conditions ideal for interspecific hybridizing and the F_1s which occurred as a result of crossing between these two species grew well in plantations. As seed was collected from these plantations and later generations from them, very marked segregation occurred and the overall performance of plantations based on such seed (in the absence of positive selection) declined progressively, the number of good trees becoming fewer the more the effects of deleterious segregation appeared in subsequent generations. Indeed, also in the Brazilian situation, it appears that additional species have been involved in the genetic interspecific mixture so that a great deal of variation from this source has been encountered.

A second example is provided by × *E. trabutii* which has been planted particularly in Italy and North Africa. The same problem with segregation from generations following the F_1 has been observed, although the F_1 from the two parents, *E. camaldulensis* and *E. botryoides* is probably better than either parent from a growth and economic point of view. The original crossing took place apparently spontaneously in Sardinia and plants, perhaps F_1s or progeny from them, were planted in the grounds of the monastery of Trefontana in Rome. It was by seed from these trees that plantations were often subsequently established. Thus as subculturing continued the increasingly marked segregation led to the same general decline in productivity which has been seen in Brazil. In many countries there are examples of such genetic degeneration, though perhaps not always so strikingly evident. One of the silvicultural needs is to prevent this by obtaining seed for planation use from sources which are genetically suitably isolated—a situation which it is not difficult to achieve by one of several means. The rate of genetic decline has been accelerated by the attraction which F_1 generation populations often have in cultivation. In spite of the large number of eucalypts, some interspecific combinations are better in specific localities in a number of desirable silvicultural features in the F_1 generation than in either genetically pure parent alone. Thus a future practice in the silviculture of eucalypts may well be the use of F_1 seed deliberately produced to gain the benefits in cultivation which follow from the first stage of interspecific hybridizing.

11.4 General uses

The biggest single urge to plant *Eucalyptus* currently in large-scale plantations is provided by the demand for wood fibre for industry. This has led to very widespread planting of *E. grandis* in areas of suitable climate, soils and topography as in South America, India and eastern and southern Africa. But in many parts of the world where there is a wood famine timber for fuel and building poles is particularly important and several eucalypt species have provided, and will continue to provide, much needed material of this kind. For example, *E. camaldulensis* is one of the most widely planted species and for such use its close relative *E. tereticornis* is exceedingly widely planted in India for both industrial wood and more direct, simpler uses. In cooler latitudes or higher altitudes *E. globulus* has been extensively planted and is a favoured producer of industrial wood in Portugal, Spain and elsewhere. On calcareous and rather low winter rainfall sites species such as *E. occidentalis* and *E. gomphocephala* have been widely used and in some of the least attractive sites for tree growing in the dry tropics *E. microtheca* has been the best performer.

At present larger markets do not favour species with wood in the higher density range but where there is not such limitation *E. citriodora*, *E. paniculata*, *E. decepta* and *E. microcorys* often perform excellently in summer rainfall regions and *E. sideroxylon*, *E. leucoxylon*, *E. gomphocephala* and *E. cladocalyx* in the winter rainfall zones.

In spite of the fact that there are such a large number of eucalypt species, those which have been widely planted and which meet most needs are relatively few and most have already been mentioned. In limited areas occasionally other species are favoured, for example *E. diversicolor* has produced some very good stands in the Cape Province of South Africa and in the central areas like Viña del Mar in Chile. *E. regnans* has grown well in both higher elevations in East Africa and southern Brazil. At times *E. viminalis* is favoured as the most useful species, as in north-east Argentina and southern Brazil.

In addition to these species some attention has been given to planting *E. astringens*, *E. nitens*, *E. elata*, *E. macarthurii*, *E. cinerea* and *E. maidenii*.

Many are grown as ornamental trees and for shade and shelter. For example, some of those most valued are *E. ficifolia* because of its brilliantly coloured flowers—vermilion, pink or orange. And other species from Western Australia like *E. erythrocorys*, *E. macrocarpa*, *E. erythronema*, *E. tetragona*, *E. torquata* are planted either because of their flowers or attractive form.

In experimental plots in arboreta and botanic gardens, most eucalypt species have been planted somewhere in the world outside Australia, but for regular use either for utility or amenity the most commonly planted are limited to some thirty or forty species.

Appendix

Eucalyptus species most commonly planted outside Australia

Species	*Climatic and soil notes*	*Geographic location*
astringens	winter rainfall, non-calcareous soil	Morocco, Tunisia, Algeria, Cyprus
†*camaldulensis*	diverse	very widespread in many countries
cinerea	cold winters	U.S.S.R. Black Sea coast, Colombia, Argentina
†*citriodora*	summer rainfall	worldwide, warm monsoon tropics
cladocalyx	winter rainfall	North Africa, South Africa
dalrympleana	cold winters	Southern Europe (France, Spain)
deglupta	wet lowland tropics	Mindanao, Papua New Guinea
diversicolor	winter rainfall	South Africa, Chile
dives	cool temperate	South Africa
ficifolia	winter rainfall	Mediterranean, South Africa, Chile
†*globulus*	Atlantic climate	Spain, Portugal, Italy, Chile, Bolivia, Nilgiri Hills (India)
**gomphocephala*	winter rainfall, calcareous soils	North Africa, Israel, Cyprus
†*grandis*	summer rainfall—equable	South and East Africa, Brazil, Kerala (India)
gunnii	cold temperate	Southern Europe, U.S.S.R. Black Sea coast
lehmannii	winter rainfall, littoral conditions	South Africa, Tunisia
macarthurii	cool temperate	South Africa (SE. Transvaal)
maculata	summer rainfall, equable conditions	South America, South Africa
maidenii	summer rainfall, equable conditions	East Africa
melanophloia	low summer rainfall	Pakistan
microcorys	equable summer rainfall	Africa, South America, Asia
†*microtheca*	low summer and winter rainfall, calcareous and gypseous soils	Sudan, North Africa, Pakistan
**occidentalis*	winter rainfall, swampy and mildly saline soils	Morocco, Algeria, Tunisia, Libya, Cyprus, Israel
**paniculata*	equable summer rainfall	East and South Africa, Brazil, India
pulverulenta	winter rainfall	U.S.A. (California)

Species	Climatic and soil notes	Geographic location
†*robusta*	summer rainfall	worldwide in tropics
rudis	winter rainfall	Pakistan, Iraq, North Africa
**sideroxylon*	winter rainfall	North Africa
†*tereticornis*	summer rainfall	India, Zambia, Rhodesia, Argentina
viminalis	equable cool temperate	Southern Brazil, Argentina, Tunisia, Italy, U.S.S.R. Black Sea coast

Note: × *Eucalyptus trabutii* is a hybrid between *E. camaldulensis* and *E. botryoides*. It has been planted widely in the Mediterranean.
'Mysore Hybrid' very extensively planted in India is *E. tereticornis*.
'*E. alba*' as used in Brazil refers to *E. 'urophylla'* and to hybrids of this species and other eucalypts—mainly *E. tereticornis*.
'*E. saligna*' as used in South and East Africa is applied very widely and incorrectly to *E. grandis*. There is little planting of true *E. saligna*.

* = Extensively planted. † = Very extensively planted.

References

ANDERSON, E. (1953). Introgressive hybridisation. *Biol. Rev.*, **28**, 28–307.

ASTON, M. J. and PATON, D. M. (1973). Frost room design for radiation frost studies in *Eucalyptus*. *Aust. J. Bot.*, **21**, 2:1–7.

BARBER, H. N. and JACKSON, W. D. (1957). Natural selection in action in *Eucalyptus*. *Nature*, **179**, 1267–9.

BENTHAM, G. (1867). *Flora Australiensis*. Vol. 3, Reeve, London.

BLAKE, S. T. (1953). Studies on northern Australian species of *Eucalyptus*. *Aust. J. Bot.*, **1**, 185–352.

BLAKELY, W. F. (1934) *A Key to the Eucalypts*, Workers Trustees, Sydney.

BLAKELY, W. F. (1965). *A Key to the Eucalypts*. 3rd edn., Commonwealth of Australia Forestry and Timber Bureau: Canberra.

BOOMSMA, C. D. (1969). Contributions to the records of *Eucalyptus* L'Héritier in South Australia. *Trans. Roy. Soc. S. Aust.*, **93**, 157–62.

BRETT, R. G. (1938). A survey of *Eucalyptus* species in Tasmania. *Pap. Proc. Roy. Soc. Tasmania*, **1937**, 75–109.

BROOKER, M. I. H. (1968). Phyllotaxis in *Eucalyptus socialis* F. Muell. and *E. oleosa* F. Muell. *Aust. J. Bot.*, **16**, 455–68.

CARR, D. J. and CARR, STELLA G. M. (1959b). Floral morphology and the taxonomy of *Eucalyptus*. *Nature*, **184**, 1549–52.

CHATTAWAY, M. MARGARET (1953). The anatomy of bark. I. The genus *Eucalyptus*. *Aust. J. Bot.*, **1**, 402–33.

CHILVERS, G. A. (1968a). Some distinctive types of eucalypt mycorrhiza. *Aust. J. Bot.*, **16**, 49.

COOLING, E. N. and ENDEAN, F. (1966). Forest Department, Zambia. *Forest Research Bulletin*, **10**.

GAUBA, E. and PRYOR, L. D. (1958). Seed coat anatomy and taxonomy in *Eucalyptus*. 1. *Proc. Linn. Soc. N.S.W.*, **83**, 20–31.

GRIEVE, B. J. (1956). Studies in the water relations of plants; transpiration of Western Australian (Swan Plain) sclerophylls. *J. Roy. Soc. Western Aust.*, **40**, 15–30.

HILLIS, W. E. (1966a). Variation in polyphenol composition within species of *Eucalyptus* L'Hérit. *Phytochemistry*, **5**, 541–56.

HUXLEY, J. (1940). *The New Systematics*. Oxford University Press, London.

JACKSON, W. D. (1958). Natural hybrids in eucalypts, Part I. *E.* × *taeniola* (=*E. salicifolia* × *sieberiana*). *Pap. Proc. Roy. Soc. Tasmania*, **92**, 141–6.

JACOBS, M. R. (1955). Growth Habits of the Eucalypts. Commonwealth of Australia Forestry and Timber Bureau, Canberra.

JOHNSON, L. A. S. (1970). Rainbow's end: the quest for an optimal taxonomy. *Syst. Zool.*, **19**, 203–39. (Reprint of JOHNSON, 1968, with addendum.)

JOHNSTON, R. D. and MARRYATT, ROSEMARY (1962). *Taxonomy and Nomenclature of Eucalypts*. Commonwealth of Australia Forestry and Timber Bureau Leaflet No. **92**, 1–24.

LARSEN, E. (1965). A Study of the Variability of *Eucalyptus maculata* Hook. and

Eucalyptus citriodora Hook. Commonwealth of Australia Forestry and Timber Bureau Leaflet No. **95**, 1–23.

MAIDEN, J. H. (1903–1931). A Critical Revision of the Genus *Eucalyptus*. Vols. 1–8, Government Printer, Sydney.

MCKERN, H. H. G. (1965). Volatile oils and plant taxonomy. (Presidential address) *J. Roy. Soc. N.S.W.*, **98**, 1–10.

MOGGI, G. (1963). L'infiorescenza del genere *Eucalyptus* e la sua interpretazione. *Giorn. Bot. Ital.*, **70**, 1–20.

MOORE, K. M. (1970). Observations on some Australian forest insects. 24. Results from a study of the genus *Glycaspis* (Homoptera: Psyllidae). *Aust. Zoologist*, **15**, 343–76.

MOORE, C. W. E. and KERAITIS, K. (1971). Effect of nitrogen source on growth of eucalypts in sand culture. *Aust. J. Bot.*, **19**, 125.

PATON, D. M. (1972). Frost resistance in *Eucalyptus*: a new method for assessment of frost injury in altitudinal provenances of *E. viminalis*. *Aust. J. Bot.*, **20**, 127–39.

PENFOLD, A. R. and WILLIS, J. L. (1961). *The Eucalypts*. Leonard Hill, London, and Interscience Publishers, New York.

PRYOR, L. D. (1954). The inheritance of inflorescence characters in *Eucalyptus*. *Proc. Linn. Soc. N.S.W.*, **79**, 79–89.

PRYOR, L. D. (1957). Variation in snow gum (*Eucalyptus pauciflora* Sieb.) *Proc. Linn. Soc. N.S.W.*, **81**, 299–305.

PRYOR, L. D. (1959). Species distribution and association in *Eucalyptus*. Biogeography and Ecology in Australia. *Mongr. Biol.*, **8**, 461–71.

PRYOR, L. D. and JOHNSON, L. A. S. (1962). The status and significance of the hybrid *Eucalyptus marginata* Sm. × *E. megacarpa* F. Muell. *Aust. J. Bot.*, **10**, 129–33.

PRYOR, L. D., JOHNSON, L. A. S., WHITECROSS, M. I. and MCGILLIVRAY, D. J. (1967). The perianth and the taxonomic affinities of *Eucalyptus cloëziana* F. Muell. *Aust. J. Bot.*, **15**, 145–9.

PRYOR, L. D. and BYRNE, O. R. (1969). Variation and taxonomy in *Eucalyptus camaldulensis*. *Silvae Genetica*, **18**, 57–96.

PRYOR, L. D. and KNOX, R. B. (1971). Operculum development and evolution in eucalypts. *Aust. J. Bot.*, **19**, 143–71.

PRYOR, L. D. and JOHNSON, L. A. S. (1971). *A Classification of the Eucalypts*, Australian National University, Canberra.

RENBUSS, M. A., CHILVERS, G. A. and PRYOR, L. D. (1972). Microbiology of an ashbed. *Proc. Linn. Soc. N.S.W.*, **97**, 4, 302–10.

STEBBINS, G. L. (1950). *Variation and Evolution in Plants*. Columbia University Press, New York.

UHLIG, S. K. (1968). Beitrag zum Problem der Mykorrhiza in *Eucalyptus*. *Zentralbl. f. Bakteriologie, Parasitenkunde, Infektions u. Hygiene*, II. **122**, 271–4.